Himalaya, Dynamics of a Giant 2

*In memory of distinguished professors
M. Gaetani, P. Molnar and A. Steck*

SCIENCES

Geoscience, Field Director – Yves Lagabrielle

Dynamics of the Continental Lithosphere,
Subject Head – Sylvie Leroy

Himalaya, Dynamics of a Giant 2

Tectonic Units and Structure of the Himalaya

Coordinated by
Rodolphe Cattin
Jean-Luc Epard

WILEY

First published 2023 in Great Britain and the United States by ISTE Ltd and John Wiley & Sons, Inc.

Apart from any fair dealing for the purposes of research or private study, or criticism or review, as permitted under the Copyright, Designs and Patents Act 1988, this publication may only be reproduced, stored or transmitted, in any form or by any means, with the prior permission in writing of the publishers, or in the case of reprographic reproduction in accordance with the terms and licenses issued by the CLA. Enquiries concerning reproduction outside these terms should be sent to the publishers at the undermentioned address:

ISTE Ltd
27-37 St George's Road
London SW19 4EU
UK

www.iste.co.uk

John Wiley & Sons, Inc.
111 River Street
Hoboken, NJ 07030
USA

www.wiley.com

© ISTE Ltd 2023

The rights of Rodolphe Cattin and Jean-Luc Epard to be identified as the authors of this work have been asserted by them in accordance with the Copyright, Designs and Patents Act 1988.

Any opinions, findings, and conclusions or recommendations expressed in this material are those of the author(s), contributor(s) or editor(s) and do not necessarily reflect the views of ISTE Group.

Library of Congress Control Number: 2022948342

British Library Cataloguing-in-Publication Data
A CIP record for this book is available from the British Library
ISBN 978-1-78945-130-6

ERC code:
PE10 Earth System Science
 PE10_5 Geology, tectonics, volcanology

Contents

Tributes xi
Eduardo GARZANTI, Vincent GODARD, Rodolphe CATTIN,
György HETÉNYI, Jean-Luc EPARD and Martin ROBYR

Foreword xvii
Rodolphe CATTIN and Jean-Luc EPARD

**Preface. From Research to Education: The Example of the
Seismology at School in Nepal Program** xix
György HETÉNYI and Shiba SUBEDI

Part 1. Tethyan Himalayan Sequence 1

**Chapter 1. Magmatism in the Kohistan–Ladakh Paleo-arc: Building
Continental Crust During the India–Eurasia Convergence** 3
Yann ROLLAND, Olivier REUBI and Sandeep SINGH

    1.1. Introduction 3
    1.2. Geological setting of the Kohistan–Ladakh arc 5
    1.3. Main geological contacts 9
        1.3.1. The Shyok Suture Zone 9
        1.3.2. The Indus Suture Zone 10
    1.4. Reconstruction of the arc structure and evolution 10

    1.4.1. Southern Plutonic Complex 11
    1.4.2. Chilas Complex 12
    1.4.3. Gilgit Complex and Kohistan Batholith 12
    1.4.4. Ladakh Batholith 13
1.5. Geochemistry and magmatic evolution of Kohistan–Ladakh magmas . 15
    1.5.1. Inferences for juvenile continental crust construction 17
    1.5.2. Isotopic composition, inferences on the mantle source and crustal
          assimilation, and implications for the timing of collisions 18
1.6. Tectonic reconstructions of Kohistan–Ladakh arc evolution 20
    1.6.1. Scenario 1: south-dipping subduction zone 20
    1.6.2. Scenario 2: north-dipping subduction zone 23
1.7. Conclusion 23
1.8. References 24

Chapter 2. Suture Zone 35
Julia DE SIGOYER and Jean-Luc EPARD

2.1. Introduction 35
2.2. General geological description of the ITSZ 38
2.3. The Indus suture zone of the Eastern Ladakh, the Nidar zone 38
    2.3.1. The Indus group sediments 41
    2.3.2. The Nidar ophiolite 41
    2.3.3. The accretionary wedge or oceanic mélange 44
2.4. Conclusion 48
2.5. References 50

Chapter 3. Geological Evolution of the Tethys Himalaya 55
Chiara MONTOMOLI, Jean-Luc EPARD, Eduardo GARZANTI,
Rodolfo CAROSI and Martin ROBYR

3.1. Introduction 55
3.2. The stratigraphy of the Tethys Himalaya 57
    3.2.1. The pre-Tethyan history 57
    3.2.2. The Neotethyan rift stage 59
    3.2.3. The Neotethyan drift stage 59
    3.2.4. The Paleocene–Eocene collision stage 63
3.3. Deformation of the Tethys Himalaya 63
    3.3.1. Deformation and metamorphism of the Tethys Himalaya in Dolpo
          (Western Nepal) 64
    3.3.2. Deformation and metamorphism of the Tethys Himalaya
          in Ladakh (NW India) 68

3.4. Conclusion	70
3.5. References	71

Part 2. Greater Himalayan Crystalline Complex — 81

Chapter 4. High-Pressure and Ultra-High-Pressure Units in the Himalaya — 83
Julia DE SIGOYER and Stéphane GUILLOT

4.1. Introduction	83
4.2. High pressure rocks in the suture zone (witnesses of the oceanic subduction)	84
4.2.1. The Shapi–Shergol blueschists (Ladakh)	86
4.2.2. The Shangla Blueschists (Pakistan)	86
4.2.3. The Sangsang Blueschist	86
4.2.4. The Indo-Burmese Blueschists	87
4.3. Continental high-pressure (HP) to ultra-high-pressure (UHP) metamorphism of the Indian margin (continental subduction) located next to the Indus Tsangpo Suture Zone	87
4.3.1. The Kaghan unit	88
4.3.2. The Tso Morari UHP unit	90
4.3.3. Other HP metamorphosed unit south of suture zone in the Indian continental margin	92
4.4. Oligocene–Miocene high-pressure, high-temperature metamorphism eclogite with granulite overprint far from the suture zone	94
4.5. Conclusion	95
4.6. References	99

Chapter 5. The Greater Himalayan Sequence – Tectonic, Petrographic and Kinematic Evolution of the Metamorphic Core Zone of the Himalayan Orogeny — 105
Martin ROBYR, Rodolfo CAROSI, Salvatore IACCARINO and Chiara MONTOMOLI

5.1. Introduction	105
5.2. Tectono-metamorphic evolution of the GHS in the central part of the Himalaya in Nepal	110
5.3. Tectono-metamorphic evolution of the GHS in the north-western part of the Indian Himalaya in Himachal Pradesh and Ladakh	117

5.3.1. Metamorphism and deformation in the High Himalayan Crystalline Zone of Zanskar 122
5.3.2. Timing of crustal shortening and metamorphism along the Miyar Shear Zone 125
5.3.3. Kinematic and tectonothermal evolution of the High Himalayan Crystalline Zone of Zanskar 126
5.4. Conclusion 128
5.5. References 129

Chapter 6. Oligo-Miocene Exhumation of the Metamorphic Core Zone of the Himalaya Across the Range 135
Rodolfo CAROSI, Salvatore IACCARINO, Chiara MONTOMOLI and Martin ROBYR

6.1. Introduction 135
6.2. Central Himalaya 136
6.3. North West India 145
6.4. Conclusion 151
6.5. References 151

Part 3. Lesser and Sub Himalayan Sequence 157

Chapter 7. Lithostratigraphy, Petrography and Metamorphism of the Lesser Himalayan Sequence 159
Chiara GROPPO, Franco ROLFO, Shashi TAMANG and Pietro MOSCA

7.1. Introduction 159
7.2. Lithostratigraphy and petrography 161
    7.2.1. Lower-LHS 162
    7.2.2. Upper-LHS 166
    7.2.3. Meta-igneous rocks 174
    7.2.4. Along-strike variation in the LHS lithostratigraphy 177
7.3. Metamorphism 177
    7.3.1. Lower-LHS 179
    7.3.2. Upper-LHS 180
    7.3.3. Tectonic implications 180
7.4. Conclusion 181
7.5. References 183

Chapter 8. Sedimentary and Structural Evolution of the Himalayan Foreland Basin 189
Pascale HUYGHE, Jean-Louis MUGNIER, Suchana TARAL
and Ananta Prasad GAJUREL

    8.1. Introduction 189
    8.2. Overall geometry of the outer Himalayan domain 191
        8.2.1. Foreland basin geometry 191
        8.2.2. Incorporation of the foreland basin into the range: a typical
            thin-skinned thrust belt structure 194
    8.3. The main foreland sediments features 198
        8.3.1. Present-day foothill sediments and morphology 198
        8.3.2. Sedimentary facies of the Neogene Siwalik foreland
            basin deposits 199
        8.3.3. Evolution of sources 205
        8.3.4. Evolution of environmental conditions 209
    8.4. Evolution of the outer Himalayan domain: geodynamics and external
        processes control 212
        8.4.1. Critical tectonic wedge, tectonic and surface processes
            velocity 212
        8.4.2. Processes controlling the evolution of the foreland basin 215
    8.5. Conclusion 218
    8.6. References 219

Conclusion 227
Rodolphe CATTIN and Jean-Luc EPARD

List of Authors 229

Index 233

Summaries of other volumes 235

Tributes

In the past five years, three of our friends, colleagues and mentors Maurizio Gaetani, Peter Molnar and Albrecht Steck have passed away. Their outstanding contribution to the knowledge of the Himalayan range has influenced many authors of this book. We pay tribute to them in the following paragraphs. We dedicate this book to these three exceptional professors.

The mountains of Asia, and the charm of romantic geology – A tribute to the legacy of Maurizio Gaetani (1940–2017) by Eduardo Garzanti

Student of Ardito Desio, organizer of the 1954 Italian conquest of K2 and younger colleague of Riccardo Assereto, killed by a landslide during the second Friuli earthquake of September 9, 1976, the everlasting love of Maurizio Gaetani for Asian geology began in 1962, with his Thesis fieldwork in the Alborz Mountains of Iran. During summer 1977, as Ladakh opened to foreigners, Maurizio first discovered with Alda Nicora the Cretaceous/Triassic boundary in the Zanskar Range. On August 1, 1981, we took a bus from Delhi to Lahul and crossed with horses and horsemen the Baralacha La and Phirtse La to describe the stratigraphy of the Paleozoic–Eocene succession of the Tethys Himalaya. New expeditions were led by Maurizio to Zanskar in 1984 and 1987 to reconstruct the paleogeographic history of northern India, from the newly identified Early Paleozoic Pan-African orogeny to Upper Paleozoic Neotethyan rifting and the subsequent Mesozoic passive-margin evolution terminated with early Paleogene collision with the Asian arc-trench system.

In the meanwhile, Maurizio's Karakorum adventure had begun with the 1986 expedition to the Hunza Valley. From then on, Maurizio's unique dedication to Karakorum geology is testified by 10 expeditions he led to Chitral, Wakhan, Shimshal and Shaksgam, during which every meter of the stratigraphic section from Ordovician to Cretaceous was measured to reconstruct the opening of Paleotethys and Neotethys and their subsequent closure during early and late Mesozoic orogenies. The amazing amount of work carried out during these surveys is condensed in the magnificent geological map of North Karakorum and summarized in his last paper "Blank on the Geological Map". The legacy of Maurizio Gaetani is to remind us that, no matter how much technology is involved, any scientific adventure is primarily a romantic adventure.

Maurizio Gaetani has contributed to about 50 articles published in peer review journals. Here is a list of his major contributions:

Gaetani, M. (2016). Blank on the geological map. *Rendiconti Lincei*, 27(2), 181–195.

Gaetani, M. and Garzanti, E. (1991). Multicyclic history of the northern India continental margin (northwestern Himalaya). *AAPG Bulletin*, 75(9), 427–1446.

Gaetani, M., Nicora, A., Premoli Silva, I. (1980). Uppermost Cretaceous and Paleocene in the Zanskar range (Ladakh-Himalaya). *Rivista Italiana di Paleontologia e Stratigrafia*, 86(1), 127–166.

Gaetani, M., Garzanti, E., Jadoul, F., Nicora, A., Tintori, A., Pasini, M., Khan, K.S.A. (1990). The north Karakorum side of the Central Asia geopuzzle. *Geological Society of America Bulletin*, 102(1), 54–62.

Zanchi, A. and Gaetani, M. (2011). The geology of the Karakoram range, Pakistan: The new 1: 100,000 geological map of Central-Western Karakoram. *Italian Journal of Geosciences*, 130(2), 161–262.

Always on the cutting edge and looking for new ideas to advance Earth Sciences – Peter Molnar (1943–2022) by Vincent Godard, Rodolphe Cattin and György Hetényi

In the final phase of preparing the three volumes of this book, we sadly learned of the passing of Peter Molnar. Peter was a giant of the Earth Sciences, whose contributions would be too long to list exhaustively in this tribute; he left a remarkably enduring mark on Himalaya–Tibet research.

Peter's scientific career began at a pivotal moment for Earth Sciences. He has significantly contributed to developing and applying the new paradigm of Plate Tectonics. Using innovative approaches based on tectonics, seismology, paleomagnetism and satellite imagery, Peter revolutionized our understanding of continental deformation and lithosphere behavior. Although Peter worked on a wide range of problems and geoscientific contexts, the India–Asia collision and the dynamics of the Himalaya–Tibet system have very often been at the core of his investigations. He has left a lasting imprint on research in this region, particularly by his desire to understand the mechanical processes at work during the deformation of this orogenic system.

Consistent with his comprehensive and integrated approach to geodynamic problems, Peter explored a wide range of ideas and processes related to the Himalaya–Tibet system's global role, and initiated several ideas and research directions that are still active today. One example, among many, is the study of the physical relationships and interaction of mechanisms between the development of Tibet's topography and the Southeast Asian monsoon regime. Among the research fields initiated by Peter, understanding the relationships between erosion, tectonics and climate is undoubtedly one of the most innovative and impactful for our community. Following a series of seminal articles by Peter and his colleagues, the complex interactions between the processes responsible for topographic relief creation and destruction are still actively debated in the Himalaya. Peter's research always focused on a global understanding of these processes, particularly those responsible for the variations in erosion and global sedimentary fluxes related to the evolution of Himalayan orogeny and the late Cenozoic evolution of climate. The fact that so many of these topics are still at the forefront of current research by so many groups worldwide is a major testimony to Peter Molnar's prescience on the dynamics of the Himalaya–Tibet system.

Beyond these outstanding contributions, Peter will be remembered for his great sense of humor and for having always been accessible and available to discuss new ideas with young scientists.

Peter Molnar has contributed to an impressive number of publications, some of which have been milestones in the understanding of the Tibet–Himalaya system:

Bilham, R., Gaur, V.K., Molnar, P. (2001). Himalayan seismic hazard. *Science*, 293(5534), 1442–1444.

Gan, W., Molnar, P., Zhang, P., Xiao, G., Liang, S., Zhang, K., Li, Z., Xu, K., Zhang, L. (2022). Initiation of clockwise rotation and eastward transport of southeastern Tibet inferred from deflected fault traces and GPS observations. *Geological Society of America Bulletin*, 134(5–6), 1129–1142.

Houseman, G.A., McKenzie, D.P., Molnar, P. (1981). Convective instability of a thickened boundary layer and its relevance for the thermal evolution of continental convergent belts. *Journal of Geophysical Research: Solid Earth*, 86(B7), 6115–6132.

Molnar, P. (2012). Isostasy can't be ignored. *Nature Geoscience*, 5(2), 83–83.

Molnar, P. and England, P. (1990). Late Cenozoic uplift of mountain ranges and global climate change: Chicken or egg? *Nature*, 346(6279), 29–34.

Molnar, P. and Tapponnier, P. (1975). Cenozoic Tectonics of Asia: Effects of a continental collision: Features of recent continental tectonics in Asia can be interpreted as results of the India-Eurasia collision. *Science*, 189(4201), 419–426.

Zhang, P.Z., Shen, Z., Wang, M., Gan, W., Bürgmann, R., Molnar, P., Wang, Q., Niu, Z., Sun, J., Wu, J. et al. (2004). Continuous deformation of the Tibetan Plateau from global positioning system data. *Geology*, 32(9), 809–812.

Geology of the Indian Himalaya – Albrecht Steck (1935–2021) by Jean-Luc Epard and Martin Robyr

Albrecht Steck was the main driving force behind the Himalayan geological research program led at the University of Lausanne over the last 40 years. His work has focused on the Indian Himalaya, particularly on the Mandi to Leh transect. He has directed or supervised eight doctoral theses distributed along this transect. One of the mottos of Albrecht was that any good geological work always starts with a sound geological mapping. Whether in the Alps or in the Himalaya, Albrecht's work excels by the quality of his geological maps. It is becoming a hallmark of Albrecht's work. Indeed, the numerous field missions he led or supervised in Lahul, Zanskar and Ladakh regions allowed the achievement of detailed geological maps covering a remarkable large area of the NW Indian Himalaya.

The research of Albrecht Steck is characterized by the combination of field observations (mapping, stratigraphy, structures and metamorphism) in order to decipher the geometry, kinematics and tectono-metamorphic history associated with orogenic processes. His scientific approach combining a variety of field, structural, petrographic and analytical methods is a hallmark of Albrecht's research. Two publications reflect particularly well the research works led by Albrecht Steck. The first (Steck et al. 1993) concerns a transect from the High Himalayan Crystalline of Lahul to the south to the Indus suture zone to the north; the second (Steck et al. 1998) focused on a complete geological transect through the Tethys Himalaya and the Tso Morari area. The results of these expeditions are also synthesized in a general publication of the Geology of NW Himalaya (Steck 2003). For Albrecht Steck, geology must be made in the field and out of the main touristic roads. His long-term commitment to the detailed study of the NW part of the Himalaya of India significantly contributed to a better understanding of the geology of this region. With Albrecht Steck, the Alpine and Himalayan geological community has lost an eminent researcher and a true Nature lover.

Albrecht Steck has contributed to many geological maps and articles published in peer-reviewed journals. Here are three of his major contributions:

Steck, A. (2003). Geology of the NW Indian Himalaya. *Eclogae Geologicae Helvetiae*, 96, 147–196.

Steck, A., Spring, L., Vannay, J.-C., Masson, H., Stutz, E., Bucher, H., Marchant, R., Tièche, J.-C. (1993). Geological transect across the Northwestern Himalaya in eastern Ladakh and Lahul (A model for the continental collision of India and Asia). *Eclogae Geologicae Helvetiae*, 86(1), 219–263.

Steck, A., Epard, J.-L., Vannay, J.-C., Hunziker, J., Girard, M., Morard, A., Robyr, M. (1998). Geological transect across the Tso Morari and Spiti areas: The nappe structures of the Tethys Himalaya. *Eclogae Geologicae Helvetiae*, 91, 103–121.

Foreword

Rodolphe CATTIN[1] and Jean-Luc EPARD[2]

[1] *University of Montpellier, France*
[2] *University of Lausanne, Switzerland*

The Himalaya is well known as the largest and highest mountain belt on Earth, stretching 2,500 km from the Nanga Parbat syntaxis in the northwest to the Namche Barwa syntaxis in the southeast, with peaks exceeding 8,000 m in altitude. Resulted from the ongoing collision between the India and Asia plates, the Himalaya is frequently used as the type example of a largely cylindrical mountain belt with a remarkable lateral continuity of major faults and tectonic units across strike.

Advances in geoscience over the past few decades have revealed a more complex picture for the dynamic of this giant, with open questions about the initial stages of the Himalayan building, lateral variations in its structures, variations in tectonic forcing, tectonic–climate coupling and assessment of the natural hazards affecting this area.

In this book, we present the current knowledge on the building and present-day behavior of the Himalayan range. The objective is not to be exhaustive, but to give some key elements to better understand the dynamics of this orogenic wedge. The three volumes of this book present (1) the geodynamic framework of the Himalayan range, (2) its main tectonic units and (3) its current activity. The chapters and volumes in this book are self-contained and can be read in any order. However, the three volumes are linked and provide together a self-consistent image of the Himalayan dynamic at various temporal and spatial scales.

This Volume 2, entitled *"Tectonic Units and Structures of the Himalaya"*, is mainly focused on field investigations to study the major structural units of this mountain belt. These up-to-date overviews include the Indus Yarlung Tsangpo suture zone, the Tethyan Himalaya sedimentary units, the Greater Himalaya Sequence, the Lesser Himalaya metamorphic rocks and the Sub-Himalaya Siwalik molasse basin.

This volume is coordinated by Rodolphe Cattin (University of Montpellier, France) and Jean-Luc Epard (University of Lausanne, Switzerland) with the help of the editorial team composed of Laurent Bollinger (French Alternative Energies and Atomic Energy Commission, France), György Hetényi (University of Lausanne, Switzerland), Vincent Godard (University of Aix-Marseille, France), Martin Robyr (University of Lausanne, Switzerland) and Julia de Sigoyer (University of Grenoble, France).

All royalties allocated to the authors of this book will be donated to the "Seismology at School" program (see the next pages).

Preface

From Research to Education: The Example of the Seismology at School in Nepal Program

György HETÉNYI[1] and Shiba SUBEDI[1,2]

[1] *Institute of Earth Sciences, University of Lausanne, Switzerland*
[2] *Seismology at School in Nepal, Pokhara, Nepal*

Scientific research aims at observing and understanding processes, and enriching our knowledge. But who is in charge of transferring this knowledge to society, to everyday life? Can we expect any researcher to become a company CEO, an engineer, a policy maker or a teacher? Our answer is no, not necessarily, but efforts can be made in that direction, and there are successful examples.

In the context of Himalayan geoscience research, a tremendous amount of information exists. It cannot be all simplified and all translated to local languages of the Himalaya; nevertheless, we found it essential that such knowledge transfer starts. In the aftermath of the 2015 magnitude 7.8 Gorkha earthquake, through a series of fortunate steps, we found ourselves putting down one of the bricks of knowledge transfer by initiating the Seismology at School in Nepal program. Our primary pathway choice was education: in the short-term, raising earthquake awareness and better preparedness can spread through the students to their families, relatives and acquaintanceship; in the longer-term, it is today's school students who will build the next generation of infrastructures.

The program started following a bottom-up approach, with direct cooperation with local schools in Nepal. This ensured motivated participants and direct feedback on the activities and about the needs. The program stands on two main pillars and is described in detail in Subedi et al. (2020a). First, earthquake-related topics have been synthesized and translated to Nepali, together with a series of hands-on experiments, and the local teachers have been trained so that they can teach these in their classes. Second, we have installed relatively cheap seismometers (RaspberryShake 1D) in local schools, which became part of the classroom activities and also recorded waves from earthquakes. This has sparked interest in schools, and the openly and publicly available waveform data is useful for monitoring and research as well. To more closely link these two, we have written a simple earthquake location tutorial that is feasible with typical school computers in Nepal (Subedi et al. 2021).

The program has started in Nepal in 2018; as of 2019, more than 20 schools and seismometers have been involved in the program, and the number is reaching 40 in 2022. There is measurable improvement in students' knowledge (Subedi et al. 2020b), and the feedbacks are very positive. Parallel to classical educational pathways, a series of other activities have been developed in the Seismology at School in Nepal program. Each school has received an Emergency Meeting Point sign in Nepali language. Over 6,000 stickers reminding about earthquakes have been distributed to increase awareness (see Figure 8.10 in Volume 3 – Chapter 8). An Earthquake Awareness Song has been written and composed, and became popular on YouTube (https://www.youtube.com/watch?v=ymE-lrAK0TI). We studied the Hindu religious representation and traditional beliefs about earthquakes (Subedi and Hetényi 2021). Recently, we have developed an educational card game to improve the practical preparation and reaction to earthquakes. Finally, we maintain a website with all information openly available (http://www.seismoschoolnp.org).

The program has so far run on funding that is considered to be small in the research domain, and this has covered the cost of materials and the work in Nepal. More recently, a crowd-funding campaign has been started and evolved successfully – we are very grateful to all funders and donors! In the future, the program aims at growing further, all across Nepal and hopefully all along the Himalaya. This will require more manpower and more funds. The authors of this book have generously given their consent to transfer all royalties to the Seismology at School in Nepal program thank you very much!

There is a strong similarity between this book and the Seismology at School in Nepal program: they both aim at taking research results and carrying them to non-specialists. This book is planned to be published in several languages, and to reach students and interested people around the world. The educational program in Nepal aims at bringing earthquake knowledge to those who really need it as they live in a high hazard area. Both efforts aim at increasing awareness, and, thereby, we hope and wish that their effects reach further across all society.

October 2022

References

Subedi, S. and Hetényi, G. (2021). The representation of earthquakes in Hindu religion: A literature review to improve educational communications in Nepal. *Front Commun*, 6, 668086. doi:10.3389/fcomm.2021.668086.

Subedi, S., Hetényi, G., Denton, P., Sauron, A. (2020a). Seismology at school in Nepal: A program for educational and citizen seismology through a low-cost seismic network. *Front Earth Sci*, 8, 73. doi:10.3389/feart.2020.00073.

Subedi, S., Hetényi, G., Shackleton, R. (2020b). Impact of an educational program on earthquake awareness and preparedness in Nepal. *Geosci Commun*, 3, 279–290. doi:10.5194/gc-3-279-2020.

Subedi, S., Denton, P., Michailos, K., Hetényi, G. (2021). Making seismology accessible to the public in Nepal: An earthquake location tutorial for education purposes. *Bull Nep Geol Soc*, 38, 149–162.

PART 1

Tethyan Himalayan Sequence

1
Magmatism in the Kohistan–Ladakh Paleo-arc: Building Continental Crust During the India–Eurasia Convergence

Yann ROLLAND[1], Olivier REUBI[2] and Sandeep SINGH[3]
[1]EDYTEM, Savoie Mont Blanc University, Le Bourget-du-Lac, France
[2]Institute of Earth Sciences, University of Lausanne, Switzerland
[3]Department of Earth Sciences, Indian Institute of Technology, Roorkee, India

1.1. Introduction

Petrological models of continental crust formation widely argue for a magmatic derivation from the mantle in magmatic arc settings. The geochemistry and production rate of arc magmas are linked to the geodynamic evolution of subduction zones. The thermal structure of the subduction zone, the flux of volatile released by the down-going slab, the crustal thickness of the arc and the nature of sediments entering the subduction zone are regarded as important parameters in controlling the geochemical characteristics of arc magmas (Ringwood 1974; Plank and Langmuir 1988, 1998; Grove et al. 2012). Furthermore, the tempo of arc magmatism, or magma addition rate, is thought to be linked to external forcing, notably changes in plate motion and subduction rate or to interconnected tectonic and magmatic processes within the arc crust (Paterson and Ducea 2015). Hence, long-term geochronological and geochemical records covering the lifetime of magmatic activity in arcs provide important constraints on evolving subduction processes during convergence and at the onset of collision in orogenic systems. The Kohistan–Ladakh paleo-arc is one of the best preserved and exposed arc sections worldwide (for recent reviews see Burg 2011; Petterson 2019). The plutonic and volcanic units cover ∼130 Ma of magmatic activity from 150 to 20 Ma. This time-span encompasses the evolution from an initial intra-oceanic subduction to partial melting of the thickened arc crust following the collision between the arc and the Indian and Eurasian continents. The Kohistan arc magmas represent a nearly complete record that provides insights into the processes of convergence and collision prior to and at the onset of the Himalayan Orogeny. This record provides insights into the debated tectonic models of arc formation in the Neotethys domain.

In addition to providing information on the orogenic processes, well-preserved arc sections are keys to understanding the formation of the continental crust. Arc magmatism along active margins represents the principal locus of production of new continental crust. Post-Archean continental growth predominantly results from accretion of the crustal section of island arcs during orogenic collisions, with an additional contribution from the intrusion of mantle-derived magmas in continental arcs (Taylor and White 1965; Rudnick 1995). Yet, melting of the metasomatized mantle wedge in subduction zones yields predominantly basaltic magmas, when estimates of the continental crust bulk composition are andesitic (Rudnick and Gao 2003). This paradox requires a mechanism removing mafic materials or adding silicic materials to the crust. In addition, the poorly understood and debated compositional stratification of the continental crust with a basaltic lower crust and a granitic upper crust requires a mechanism of differentiation within the arc crust. To address the formation and evolution of continental crust, it is consequently essential to characterize the compositional architecture of crustal arc sections, to document the mechanisms controlling magma differentiation in the crust and to establish the processes controlling mass addition or removal from the evolving crust. The Kohistan–Ladakh plutonic and volcanic units form a nearly complete crustal section from upper-mantle to lower-crustal ultra-mafic rocks through mid-crustal batholiths and to the volcanic cover, a nearly unique occurrence in the geological record (Jagoutz et al. 2009, 2011; Jagoutz 2010). This preserved crustal section (Figure 1.1) offers an exceptional opportunity to document the locus and mechanisms of differentiation in arc crustal sections.

The objectives of this chapter are to review the general geological, petrological and geochemical characteristics of the Kohistan–Ladakh rocks and discuss the implications for the Himalayan Orogeny, as well as for the production of continental crust through arc magmatism. Among arising questions concerned by this review are (i) the duration of magmatic arc construction; (ii) are there one or several arcs? and (iii) the timing of arc accretion to India and Asia, and insights for tectonic reconstructions of India–Asia convergence models (see Volume 1 – Chapter 1). This chapter summarizes the possible reconstructions based on recent data.

1.2. Geological setting of the Kohistan–Ladakh arc

The Kohistan–Ladakh arc forms a coherent tectonic block within the Himalayan belt, although the Kohistan and Ladakh sections are separated by the Nanga Parbat syntaxis (Figures 1.1 and 1.2).

It is considered to be part of the Trans-Himalayan plutonic complex, a 2,500 km linear suite of plutonic units extending from Pakistan in the west to Tibet in the east (Honegger et al. 1982; Bard 1983; Rolland et al. 2000). The Kohistan–Ladakh arc is delimited to the north by the Shyok (or Karakoram–Kohistan) suture and to the south by the Indus Suture (Figure 1.1). Eastward, the Shyok suture is dissected by the Karakoram fault and gives place to the Tsangpo suture delimitating India and the Eurasian margin (Figure 1.1). To the south, the Indus Suture marks the limit with the Indian plate. The suture contains ophiolitic sequences and ultra-high pressure rocks remnants of a subduction zone (Figure 1.2).

The Kohistan–Ladakh paleo-island arc developed during the Mesozoic in the equatorial area of the Neotethys Ocean, south of Asia and was wedged between the Indian and Asian plates during the Himalayan Orogeny (Bard 1983; Khan et al. 2009; Burg 2011; Bouilhol et al. 2013; Martin et al. 2020). U–Pb zircon ages indicate that the magmatic activity in the KLA started approximately 150 Ma and lasted until 20 Ma, with two main episodes of subduction-related magmatism (120–80 Ma and 80–40 Ma) (Figure 1.3) (for recent age compilations see Bouilhol et al. (2013), Burg and Bouilhol (2019) and Jagoutz et al. (2019)). North of the Shyok suture, a second subduction system was active simultaneously on the Eurasian continental margin and resulted in the formation of the Karakoram Batholith with U–Pb zircon ages ranging from 120 to 40 Ma (e.g. Debon et al. 1987).

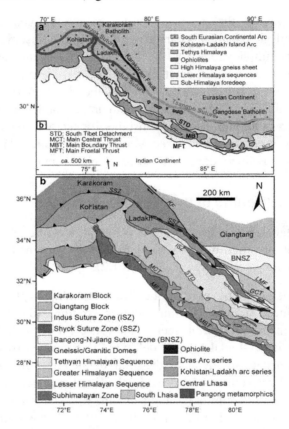

Figure 1.1. *(a) Location of Kohistan–Ladakh and ophiolites in the Himalayan Belt. (b) Simplified geological map of the western Himalayan orogen (modified after Parsons et al. 2020). Tectonic structures: ATF=Altyn Tagh Fault; BNSZ=Bangong–Nujiang Suture Zone; GCT=Great Counter Thrust; ISZ=Indus Suture Zone; JSZ=Jinsha Suture Zone; KF=Karakoram Fault; KSZ=Kunlun Suture Zone; LMF=Luobadui–Milashan Fault; MCT=Main Central Thrust; MFT=Main Frontal Thrust; SSZ=Shyok Suture Zone; STD=South Tibetan Detachment. (a) Location of Kohistan–Ladakh and ophiolites in the Himalayan Belt. (b) Simplified geological map of the western Himalayan orogen (modified after Parsons et al. 2020). Tectonic structures: ATF=Altyn Tagh Fault; BNSZ=Bangong–Nujiang Suture Zone; GCT=Great Counter Thrust; ISZ=Indus Suture Zone; JSZ=Jinsha Suture Zone; KF=Karakoram Fault; KSZ=Kunlun Suture Zone; LMF=Luobadui–Milashan Fault; MCT=Main Central Thrust; MFT=Main Frontal Thrust; SSZ=Shyok Suture Zone; STD=South Tibetan Detachment. For a color version of this figure, see www.iste.co.uk/cattin/himalaya2.zip*

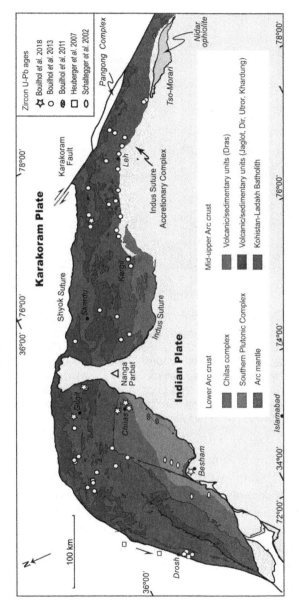

Figure 1.2. *Geological map of the Kohistan–Ladakh arc (modified from Jagoutz et al. 2019). Symbols indicate samples for which whole-rock geochemistry and zircon U–Pb ages were obtained, for a synthesis (see Jagoutz and Behn 2013). For a color version of this figure, see www.iste.co.uk/cattin/himalaya2.zip*

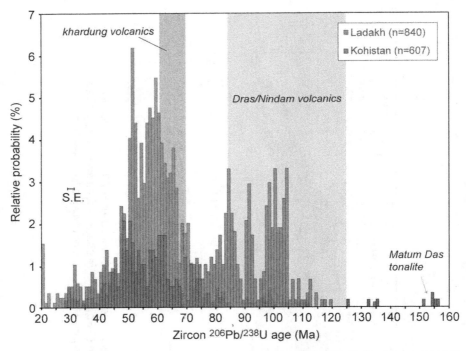

Figure 1.3. *Relative distribution of U–Pb Zircon ages from the Kohistan and Ladakh plutonic sections of the Kohistan–Ladakh arc (data are from Bouilhol et al. 2011, 2013; Bosch et al. 2011; Heuberger et al. 2007; Jagoutz et al. 2019; Ravikant et al. 2009; Schaltegger et al. 2002; Upadhyay et al. 2008). The shaded areas show the U–Pb Zircon age ranges for the Khardung volcanics, north Ladakh (Lakhan et al. 2020) and volcaniclastic Nindam formation, south Ladakh (Walsh et al. 2019). The latter is believed to represent the distal turbidities from the Dras arc. For a color version of this figure, see www.iste.co.uk/cattin/himalaya2.zip*

East from the Karakoram fault, the Karakoram Batholith gives place to the Gangdese Batholith. Timing of the collision of the arc(s) system(s) with the Eurasian and Indian continents is still contentious. Closure of the Shyok suture has been inferred to either predate or postdate the collision between KLA and India (Petterson and Windley 1985; Khan et al. 2009; Bouilhol et al. 2013; Walsh et al. 2019; Martin et al. 2020). In addition, it is debated whether part of the Ladakh sector of the arc developed on a continental or oceanic basement (e.g. Rolland et al. 2000; Burg 2011). These points are discussed further below.

1.3. Main geological contacts

1.3.1. *The Shyok Suture Zone*

The Shyok Suture Zone separates the Kohistan–Ladakh block from the Karakoram margin of Eurasia (Coward et al. 1986; Pudsey 1986; Rolland et al. 2000). The latter records Paleozoic (Ordovician)–Mesozoic sedimentation (Gaetani 1997; Rolland et al. 2002) on a >651 Ma Precambrian basement (Rolland et al. 2006b). Several phases of magmatism are recorded since the Upper Triassic to the Miocene (Rolland et al. 2001, 2006b; Heuberger et al. 2007; Ravikant et al. 2009; St-Onge et al. 2010; Phillips et al. 2013; Borneman et al. 2015; Kumar et al. 2017); Lower to Upper Jurassic and Lower to Upper Cretaceous phases of deformation (Gaetani et al. 1990), and middle Cretaceous to Cenozoic metamorphic events (Rolland et al. 2006a; Searle and Hacker 2019). To the East, the Qiangtang block might represent a lateral equivalent of the Karakoram across the Karakoram Fault, and the Bangong suture is considered equivalent to the Shyok suture (Phillips et al. 2004; Baxter et al. 2009; Robinson 2009; Rolland et al. 2009; Borneman et al. 2015). While the Bangong suture is generally considered to be Cretaceous in age, the Shyok suture closure age is more controversial. Geologic, metamorphic and stratigraphic data from the Shyok suture and adjacent Karakoram block support a middle Cretaceous accretion of the Kohistan–Ladakh arc (Rex et al. 1988; Treloar et al. 1989; Rolland et al. 2000; Palin et al. 2012; Borneman et al. 2015; Kumar et al. 2017), whereas geochemical and paleomagnetic data from the Kohistan–Ladakh arc are possibly indicative of a collision with India at ca. 50 Ma and a later collision with Karakoram at ca. 40 Ma (Bouilhol et al. 2013; Martin et al. 2020). However, age relationships provided by Borneman et al. (2015) are clearly indicative of a minimum Late Cretaceous age for the Shyok Suture. These authors showed that the youngest population of detrital zircons obtained in the overlying Saltoro molasse rocks is ~92 Ma and obtained a U/Pb zircon date for a dike that cuts the basal molasse outcrops (~85 Ma), which clearly rules out the possibility of an Eocene collision between Kohistan–Ladakh and Eurasia.

The eastward correlation of the Kohistan–Ladakh and Dras arc remnants with Tibetan units is also controversial, because a clear lateral continuation is hindered by the offset of the Karakoram Fault (Figure 1.1). Although built on a basement of distinct nature, these arc remnants are commonly correlated with those of the Lhasa block, which consists of a Precambrian basement overlain and intruded by Paleozoic to Cenozoic strata and magmatic rocks (Dürr 1996; Zhu et al. 2009). This correlation is based on the similar age, structural position and geochemistry of the Upper Cretaceous–Cenozoic batholiths from both arcs (Gangdese, Ladakh, Kohistan; Raz and Honegger 1989; Rolland et al. 2000; Weinberg et al. 2000; Ji et al. 2009). Alternative connections were proposed between intra-oceanic arc remnants located south of Lhasa (Zedong arc; Aitchison et al. 2000, 2007; McDermid et al. 2002) and either the Kohistan–Ladakh arc (Parsons et al. 2020), the Dras arc (e.g. Walsh et al. 2021) or the Spong arc (e.g. Hébert et al. 2012).

1.3.2. The Indus Suture Zone

The Indus Suture Zone in NW Ladakh is composed of igneous, metamorphic and sedimentary units (Figure 1.2). Here, a blueschist facies suture is exhibited (Honegger et al. 1982). Exposed north of the suture zone are the intrusive and volcanic units of the Kohistan–Ladakh arc. South of the suture, the Indus Group overlies the middle Cretaceous to lower Eocene Tar Group, which consists of the Aptian–Albian Khalsi Limestone (Garzanti and Van Haver 1988; Searle et al. 1990; Clift et al. 2000; Green et al. 2008; Henderson et al. 2010). The Tar Group is generally considered as the forearc basin of the Kohistan–Ladakh arc (Garzanti and Van Haver 1988; Searle et al. 1990; Green et al. 2008; Henderson et al. 2010). The Upper Jurassic to Paleocene Dras intraoceanic arc is tectonically juxtaposed against the Ladakh Batholith, the Tar Group and the Indus Group, to the north, and thrust over the Indian passive margin, to the south. The western part of the Dras arc is dominated by mafic to intermediate volcanic rocks, which transition eastwards into the deep-water volcaniclastic rocks of the Aptian to Paleocene Nindam Formation (Honegger et al. 1982; Dietrich et al. 1983; Reuber 1989; Clift et al. 2000; Bhat et al. 2019; Walsh et al. 2021). Exposed south of the suture zone is the Permian to lower Eocene Zanskar Supergroup of the Tethyan Himalaya (northern Indian margin), which is thrusted by the Jurassic to Upper Cretaceous Spongtang supra-subduction zone ophiolite (e.g. Reuber et al. 1987; Corfield and Searle 2000; Mahéo et al. 2004).

1.4. Reconstruction of the arc structure and evolution

The Kohistan–Ladakh paleo-arc is composed of several major plutonic-volcanic sections (Figure 1.2). The zonality of magmatic sections shows a progressive evolution from lower crustal cumulates in Kohistan towards mid-crustal plutons in both central Kohistan and in Ladakh, and more superficial intrusives and volcanic complexes on the northern side of Kohistan–Ladakh, along the Shyok suture.

The Kohistan sequence makes up the most complete arc section and is thus presented in detail in this article. The Kohistan section is formed by three major units from south to north: the Southern Plutonic Complex, the Chilas Complex and the Gilgit Complex (Figure 1.2). Detailed descriptions of these complexes can be found in Tahirkheli (1979), Jan and Howie (1980), Bard (1983), Petterson and Windley (1985), Burg et al. (1998), Petterson and Treloar (2004), Yoshino and Okudaira (2004), Jagoutz et al. (2006), Garrido et al. (2007), Jagoutz et al. (2007), Dhuime et al. (2007), Bouilhol et al. (2009), Dhuime et al. (2009), Jagoutz (2010), Burg (2011) and Petterson (2019). The inferred crustal thicknesses of the Kohistan arc ($\sim$54–55 km) is significantly more than any currently active intra-oceanic arc, which have a crustal thickness of about 35 km. The Ladakh section is regarded as a lateral equivalent of the Gilgit Complex and represents an inferred crustal thickness of $\sim$30 km.

1.4.1. *Southern Plutonic Complex*

The Southern Plutonic Complex represents the lower arc crust and mantle transition, mainly formed of ultramafic rocks exposed in the Jijal area along the Indus River (Figure 1.2; Jan and Howie 1980; Miller et al. 1991). At its base (Jijal and Sapat sections), it comprises ultramafic to mafic rocks composed predominantly of clinopyroxene-rich lithologies (pyroxenite, wherlite and websterite) with subordinate harzburgite, dunite and garnet-gabbros. Amphibole becomes more important up section. Despite the granulite grade metamorphism, original igneous textures and structures are prevalent. The upper section of the complex (Kamila or Southern Amphibolite) consists of a 15–35 km thick pile of sheet-like intrusions composed of gabbros and diorites with subordinate tonalities, granites and trondhjemites. The Southern Plutonic Complex is characterized by originally horizontal sheet-like and lens-like intrusions. The mineral assemblages indicate pressures ranging from 1.5–1.8 GPa at the base of the complex to 0.7 GPa at the top. Zircon

U–Pb ages on zircons indicate that the complex was formed from ∼118 to 85 Ma (Schaltegger et al. 2002; Dhuime et al. 2007; Bosch et al. 2011).

1.4.2. Chilas Complex

The mid-crustal Chilas Complex is mainly composed of gabbro(-norite) associated with mafic–ultramafic intrusions and minor diorites (Jagoutz et al. 2006, 2007). Tabular bodies of ultramafic rocks are sparce and are dominated by dunite with minor websterite and pyroxenite. Amphibole is generally absent or appears as a late magmatic phase. The mineral assemblage records formation pressures of ∼7 kbar (Jagoutz et al. 2007). The Chilas Complex formed at ∼85 Ma (Schaltegger et al. 2002). It is believed to have formed during intra-arc extension and to represent a relatively dry fractionation sequence distinct from the wet differentiation sequence recorded by the Southern Plutonic and Gilgit Complex (Jagoutz et al. 2011).

1.4.3. Gilgit Complex and Kohistan Batholith

The Gilgit Complex represents the mid- to upper arc crust, and is dominated by the calc-alkaline Kohistan Batholith, which makes up the northern two-thirds of the Kohistan block, and is equivalent to most of the Ladakh side. The volcanic units range from basalt to rhyolite (Petterson and Treloar 2004), while mafic basaltic to andesitic compositions predominate, alternating with marine volcanic breccia and limestones representing a marginal basin arc environment, like in northern Ladakh (Rolland et al. 2000).

The Kohistan Batholith range in compositions from rare gabbro through diorite, tonalite, granodiorite and granite. Amphibole is present in most batholith rocks, when pyroxene is rare. The batholith comprises numerous individual plutons, stocks, sheets and dykes ranging in size from a few meters to 5–15 km in diameter. Many plutons are compositionally heterogeneous at outcrop scale (Petterson and Windley 1985). Diorite dykes are common in the deeper part of the batholith in the south, when aplite-pegmatite to leucogranite sheets are abundantly close to the Nanga Parbat syntaxis (Petterson and Windley 1985; Jagoutz and Schmidt 2012). Emplacement pressure estimates based on amphibole barometry increase from < 2 kbar in the NW to ∼8 kbar in the SE and are influenced by the presence of the Nanga Parbat syntaxis

(Jagoutz 2014). U–Pb zircon ages indicate that the magmatic activity began at least ca. 112 Ma and lasted ca. 28 Ma, with the bulk of the batholith emplaced between 85 and 40 Ma (Jagoutz et al. 2009; Bouilhol et al. 2013).

1.4.4. Ladakh Batholith

The Ladakh Batholith is composed of calc-alkaline intrusions with compositions varying from noritic gabbro to granite (Honegger et al. 1982; Debon et al. 1987; Ahmad and Islam 1998; Rolland et al. 2000; Weinberg and Dunlap 2000; Singh et al. 2007; White et al. 2011). Vast hornblende-biotite granodiorite occurs as the main rock type with regional strike of NW-SE and steep dips towards north. Remnants of dioritic and gabbroic bodies intruded by the granodiorite are seen in the southern parts while ascending to Khardung La. Aplitic veins intruding the main mass are common, while numerous dolerite dikes cut across the granodioritic mass and show chilled contact. Light colored exposures of granite and leucogranite are also observed locally. The Ladakh Batholith is largely undeformed. A diffuse deformation zone with an intense penetrative ductile shear fabric has nevertheless been observed by Weinberg and Dunlap (2000) within the NNW-trending dextral Thanglaso Shear Zone north of Leh. Biotite-rich enclaves that are possible remnants of biotite schist are found in alternation with marble in the central part of the batholith (Rolland et al. 2000), and may be remnants of volcano-sedimentary series. Based on different geochronometers, the batholith dates back to as early as 102 ± 2 Ma (Schärer et al. 1984) with the most recent magmatic activity to approximately 46 Ma (Figure 1.3) (Honegger et al. 1982; Weinberg and Dunlap 2000; White et al. 2011). Emplacement pressure estimates based on amphibole barometry increase from 2 kbar in the north to $\sim$7 kbar in the south (Singh et al. 2007; Reubi and Müntener 2022).

To the north, the Ladakh Batholith shows unconformable relationships with the younger (Cenozoic) Khardung Volcanics, which are mainly rhyolite, andesite, agglomerate, gravelly sandstone and conglomerate intercalations (Rolland et al. 2000; Lakhan et al. 2020). To the southwest, the batholith emplaced into an unmetamorphosed thick pile of basic and acid volcanics (Dras Volcanics). To the southeast, it is covered by a vast alluvium of the Indus River.

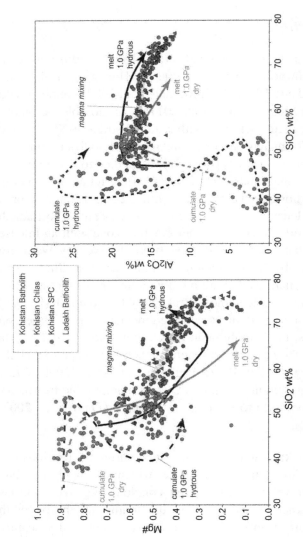

Figure 1.4. *Harker diagrams illustrating the different fractionation trends for the plutonic rocks from the Southern Plutonic (SPC), Chilas and Batholith complexes in Kohistan and the Ladakh Batholith. The black arrow represents the compositional evolution of the melts from hydrous fractional crystallization experiments at 1.0 GPa (data from Ulmer et al. 2018), when the dashed black arrow represents the evolution of the cumulates in these experiments. The gray arrow shows the melt evolution and the gray dashed arrow the cumulate evolution in anhydrous fractional crystallization experiments at 1.0 GPa (data from Villiger et al. 2004). The dashed yellow double arrow shows an arbitrary mixing trends producing the andesitic magmas through mixing between the mafic and silicic magmas produced by high-pressure hydrous fractional crystallization. For a color version of this figure, see www.iste.co.uk/cattin/himalaya2.zip*

1.5. Geochemistry and magmatic evolution of Kohistan–Ladakh magmas

According to their texture and geochemistry, plutonic rocks can be divided into two broad types; plutonic rocks with near melt chemical compositions and cumulates formed by the accumulation of mineral with some remaining melt. The plutonic rocks from the Southern Plutonic Complex mostly represent cumulate compositions, when the plutonic rocks from the Kohistan–Ladakh Batholith are, for the majority, representative of melt compositions (Figures 1.4 and 1.5) (Jagoutz et al. 2011; Jagoutz 2014).

The melt representative plutonic rocks follow a calc-alkaline fractionation trend, and incompatible trace elements are enriched compared to average mid-ocean ridge basalts (MORB) and depleted in high field strength elements (HFSE) compared to large ion lithophile elements (LILE), as typical for subduction zone magmatism. Based on their geochemistry, the melt representative plutonic rocks can be divided into two differentiation series: those from the Chilas Complex are characterized by a more drastic increase in incompatible elements (e.g. K_2O) over a given silica-enrichment compared to the Kohistan–Ladakh Batholith rocks (Figure 1.6) (Jagoutz et al. 2011). A comparison with melt compositions obtained experimentally demonstrates that these series results from distinct high-pressure fractionation trends, a relatively dry fractionation sequence of olivine–clinopyroxene–orthopyroxene–plagioclase–amphibole in the Chilas case, and a hydrous sequence of olivine–pyroxenes–garnet–Fe/Ti–oxide–amphibole–plagioclase for the Kohistan–Ladakh Batholith (Figure 1.4) (Jagoutz et al. 2011).

Furthermore, the Southern Plutonic Complex cumulates are mineralogically and geochemically similar to the cumulates produced experimentally by the fractional crystallization of hydrous mafic melts at high pressure, when the Chilas cumulates are equivalent to cumulates of dry high-pressure experiments (Figure 1.4) (Müntener and Ulmer 2018; Jagoutz et al. 2019).

These observations support the inference that the Chilas magmas were produced by decompression-melting during intra-arc extension and represent a relatively dry fractionation sequence distinct from the predominant wet differentiation sequence recorded by the Southern Plutonic Complex and the

Kohistan–Ladakh Batholith (Jagoutz et al. 2011). In addition, this indicates that the cumulates from Southern Plutonic Complex are complementary to the differentiated magmas from the batholith. Overall, the Kohistan magmatic arc section demonstrates that the bulk of differentiation that produced the silicic magma ubiquitous at mid- to upper-crust depths occurred predominantly by fractional crystallization in the lower crust ($\geq$ 30 km) (Figure 1.5). The geochemistry of batholith rocks requires nevertheless that a lesser amount of differentiation occurred as these magmas were ascending through the mid- to upper-crust (Reubi and Müntener 2022).

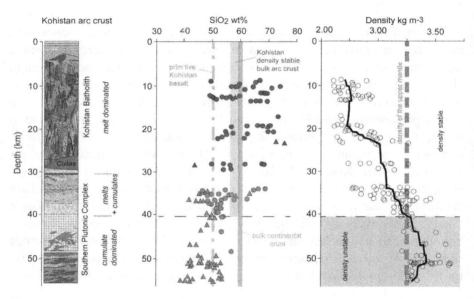

Figure 1.5. Variation of bulk-rock SiO_2 contents and melt or cumulate density versus intrusion depth inferred from amphibole barometry pressure [modified from Jagoutz and Behn (2013)] and Jagoutz (2014). Triangles indicate plutonic rocks with near liquid chemical composition, and circles indicate cumulate rocks. The gray dashed line indicates the average composition of Kohistan primitive arc volcanics (after Jagoutz 2010). The light red box represents the volumetrically pondered average composition estimated for the density stable section of the Kohistan arc crust (after Jagoutz and Schmidt 2012). The orange line shows the compositional estimate for the bulk continental crust from Rudnick and Gao (2003). The details for the melt/cumulate density calculation can be found in Jagoutz and Behn (2013). For a color version of this figure, see www.iste.co.uk/cattin/himalaya2.zip

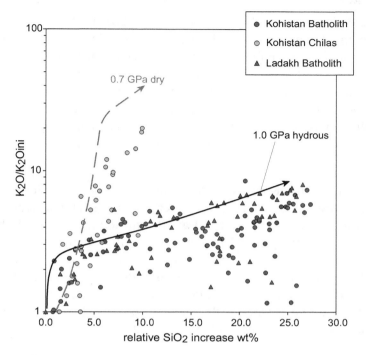

Figure 1.6. *Compositional evolution of the Chilas and Kohistan–Ladakh rocks representative of melt compositions compared to experimental melts from hydrous and anhydrous high-pressure fractional crystallization [experimental data from Villiger et al. (2004) and Ulmer et al. (2018). All data are normalized to the K_2O and SiO_2 contents in the least differentiated magmas/melt in each series. For a color version of this figure, see www.iste.co.uk/cattin/himalaya2.zip*

1.5.1. *Inferences for juvenile continental crust construction*

The Kohistan–Ladakh nearly complete crustal arc section provides unprecedented insights into the possible mechanisms that yield an andesitic (∼60 wt% SiO_2) continental crust in oceanic arcs. In Kohistan, like in most arcs, primitive hydrous basalts produced by melting of the metasomatized mantle wedge have a broadly basaltic composition with approximately 50 wt% SiO_2 (Jagoutz 2014; Reubi and Müntener 2022). Crystallization of these compositions in the lower crust (at pressures of 1.5–0.7 GPa) produces, on one hand, garnet- and/or amphibole-rich cumulates that have a density exceeding the density of upper mantle peridotites, and, on the other hand, less dense silicate rich melts (Figure 1.5). Foundering of the silicate-poor and gravitationally unstable cumulates in the mantle leaves by mass balance a silicate-enriched crust. (Jagoutz and Schmidt 2012) estimated the average composition of the density stable crustal section in Kohistan to have an andesitic composition similar to the bulk continental crust estimates with SiO_2 of 56.6–59.3 wt% (Figure 1.5). Foundering of lower island-arc crust appears as a plausible mechanism to produce andesitic crust from primitive basaltic melts (Jagoutz and Behn 2013). The Kohistan–Ladakh arc is thus an example of how the continental crust can be derived from mantle melting in subduction zones, and argues for a progressive construction of continental crust through geological times.

In addition, the Kohistan–Ladakh magmatic section illustrates how magmatic differentiation can yield a chemically stratified crust with a predominately mafic lower crust and a differentiated but essentially bimodal upper crust. Interestingly, andesitic magmas with a composition similar to the bulk continental crust are rare in the density stable part of the section (Figure 1.5). This warrants caution against models arguing for derivation of the continental crust by primary magmatic processes (e.g. Kelemen 1995; Hacker et al. 2011) and rather indicates the importance of mass balance between the upper and lower crusts (Reubi and Müntener 2022).

1.5.2. *Isotopic composition, inferences on the mantle source and crustal assimilation, and implications for the timing of collisions*

The isotopic compositions of the Kohistan–Ladakh arc rocks results to εSr (–4 to –6) εHf (11.4–12.5), εNd (4.8–5.0), $^{208/204}$Pb (38–39), $^{207/204}$Pb (15–16) and $^{206/204}$Pb (18–19) (Khan et al. 1997; Rolland 2002; Jagoutz and Schmidt 2012; Bouilhol et al. 2013; Jagoutz et al. 2019), which is more evolved than modern pacific intra-oceanic arcs. These isotopic compositions fall within the range of DUPAL anomaly, which led Khan et al. (1997) to propose the derivation from mantle in the South Indian Ocean, where this anomaly is present. However, this evolved isotopic composition could also represent the involvement of a crustal component such as a continental basement or <10% of pelagic sediments (Rolland 2002; Bouilhol et al. 2013; Jagoutz et al. 2019). The depleted mantle component of the source is considered to be paleo-Tethyan MORB-type mantle (Khan et al. 1997) with isotopic signatures similar to the present-day Indian MORB-type mantle (Mahoney et al. 1998). Hf isotopes and U–Pb ages from single zircons, as well as bulk-rock Nd isotopes show a steady evolution to isotopically enriched compositions (decreasing εHf and εNd values) from 154 Ma to 50 Ma (Figure 1.7).

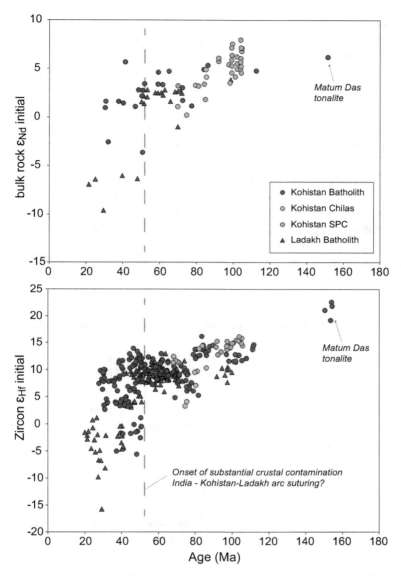

Figure 1.7. Hf isotopes from single zircons and bulk-rock Nd isotopic compositions as a function of U–Pb zircon ages of plutonic rocks from the Kohistan–Ladakh arc (data are from Schaltegger et al. 2002; Heuberger et al. 2007; Bouilhol et al. 2011, 2013; Jagoutz et al. 2019). For a color version of this figure, see www.iste.co.uk/cattin/himalaya2.zip

This temporal trend is concomitant with bulk-rock increases in εSr, $^{208/204}$Pb, $^{207/204}$Pb and $^{206/204}$Pb (Jagoutz et al. 2019), suggesting an increasing influence of the sediment component issued from the subducting slab. The implications of that still need to be assessed and could indicate either an increase in the volume of the subducted sediments, an increase in the subduction flux, an increasing influence of terrigenous sediments or changes in the thermal state of the slab. Marked shifts to negative εHf and εNd values translating substantial crustal contamination of the arc magma source are only recorded after 50 Ma (Figure 1.7). These data argue for the collision between India and the Kohistan–Ladakh arc at 50Ma (Bouilhol et al. 2013).

Based on an apparent diachronic record of the shift to substantial crustal contamination between the south and the north of the Ladakh Batholith, Bouilhol et al. (2013) further suggested that the collision between the arc and India occurred first at 50 Ma along the Indus suture and that the final collision with Eurasia along the Shyok occurred at 40 Ma.

1.6. Tectonic reconstructions of Kohistan–Ladakh arc evolution

Based on the work and discussions of Andjić et al. (2022), tectonic reconstructions are proposed for the Kohistan–Ladakh arc, which is here assumed to be continuous with the Dras arc (KLD arc). Two hypotheses can be proposed (Figures 1.8 and 1.9): either (1) a south-dipping intra-oceanic subduction zone (scenario 1) or (ii) a north-dipping intra-oceanic subduction zone (scenario 2).

1.6.1. *Scenario 1: south-dipping subduction zone*

Arc continent accretion along the Shyok suture zone occurred either during the middle Cretaceous (e.g. Rolland 2002; Borneman et al. 2015) or the Eocene (e.g. Bouilhol et al. 2013; Martin et al. 2020). A Cretaceous age for the Shyok suture is in agreement with the inferred lateral correlation between the Shyok and the Cretaceous Bangong suture in Tibet. In this scenario, these sutures were formed by the southward subduction of an ocean that was bounded to the south by the Kohistan–Ladakh–Dras arc in continuity with the Lhasa Terrane (Gangdese Batholith) (scenario 1 in Figure 1.8). A subduction onset in the Jurassic is also argued for beneath the Lhasa Terrane (e.g. Zhu et al. 2009).

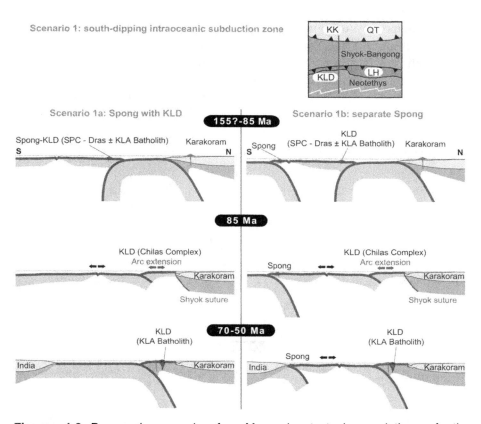

Figure 1.8. *Proposed scenario for Mesozoic tectonic evolution of the Kohistan–Ladakh–Dras (KLD) arc (after Andjić et al. 2022, modified). Scenario 1: the Kohistan–Ladakh–Dras arc formed along a south-dipping intra-oceanic subduction zone and was laterally continuous with the Lhasa Terrane (LH) to the east. The Spong arc may or may not be part of the same convergent margin as the Kohistan–Ladakh–Dras arc. 180–160 Ma=Early to Late Jurassic; 140 Ma=Early Cretaceous; 125 Ma=Early Cretaceous; 100–80 Ma=Late Cretaceous. KK=Karakoram; LH=Lhasa; SDLK=Spong–Dras–Ladakh–Kohistan; QT=Qiangtang. For a color version of this figure, see www.iste.co.uk/cattin/himalaya2.zip*

Tectonic accretion of the Kohistan–Ladakh–Dras arc to Eurasia during the Early Cretaceous is evidenced by a magmatic lull in those arcs and in the Karakoram margin (ca. 140–120 Ma) (Searle et al. 1989; Heuberger et al. 2007; Borneman et al. 2015; Searle and Hacker 2019). A possible re-start of magmatic activity was then triggered by the onset of northward subduction in the mid-cretaceous, evolving by roll-back towards intra- and back-arc basins (Rolland et al. 2000; Rolland 2002; Robertson and Collins 2002). Initiation of a southward intra-oceanic subduction along the KLD arc and away from the northern margin of India is supported by the accretion of Upper Permian seamounts to the Spong arc in post-Campanian times (Reuber et al. 1987; Corfield and Searle 1999).

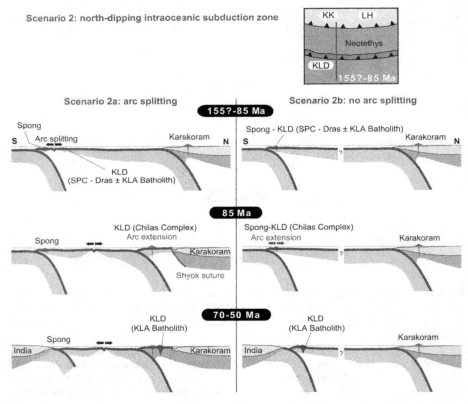

Figure 1.9. *Proposed scenario for Mesozoic tectonic evolution of the Kohistan–Ladakh–Dras (KLD) arc (after Andjić et al. 2022, modified). Scenario 2: the Kohistan–Ladakh–Dras formed along a north-dipping intra-oceanic subduction zone. Subduction initiation may have occurred as early as the Early Jurassic along a Neotethyan mid-ocean ridge. Pre-Late Jurassic accretion of Lhasa to Eurasia is assumed in scenario 2. 180–160 Ma=Early to Late Jurassic; 140 Ma=Early Cretaceous; 125 Ma=Early Cretaceous; 100–80 Ma=Late Cretaceous. KK=Karakoram; LH=Lhasa; SDLK=Spong–Dras–Ladakh–Kohistan; QT=Qiangtang. For a color version of this figure, see www.iste.co.uk/cattin/himalaya2.zip*

1.6.2. Scenario 2: north-dipping subduction zone

A simpler scenario of north-dipping subduction zones with accretion of an intra-oceanic KLD arc to the Karakoram during the Early Cretaceous may also agree with a Cretaceous age for the Shyok suture. In this scenario, the KLD arc drifted northward from a position within the Neotethys and the Lhasa Terrain had previously accreted to Eurasia (scenario 2a in Figure 1.9).

The observation that Lhasa-type and Eurasia-type zircons were transported to the Kohistan–Ladakh–Dras arc by the middle Cretaceous (Andjić et al. 2022) supports this scenario. It would require, however, a correlation of the Lhasa and Karakoram Terranes (e.g. Yang et al. 2017; Robinson 2009). Alternatively, a post-Cretaceous accretion of the KLD arc to the Karakoram Terrane along the Shyok suture would imply that the KLD arc remained in an intra-oceanic arc position until the Paleocene (e.g. Jagoutz et al. 2019; Bouilhol et al. 2013) (Figure 1.9, scenario 2b). The absence of a continent or a continental fragment in the vicinity of the KLD subduction zone makes, however, this scenario the least likely solution, because Lhasa- and Eurasia-type zircons were found together in the KLD arc by the middle Cretaceous by Andjić et al. (2022). Furthermore, the inferred Paleocene latitude of the KLD arc, a key argument to support an intra-oceanic position, is based on a single paleomagnetic pole, and is thus still ill-defined.

1.7. Conclusion

The Kohistan–Ladakh arc represents the better-preserved and most complete crustal arc section on Earth. Obduction of the arc onto the Indian margin to the south led to the uplift of the lower crustal section of the arc, while accretion to the Asian margin to the north exhibits a tectonic melange of diverse marginal arc units. This whole geological object provides valuable information on the processes of differentiation of a magmatic intra-oceanic arc and for the creation of juvenile continental crust. In addition, the Kohistan–Ladakh and Dras arcs witness the pre-collisional stages of India–Asia convergence, a necessary input for any precise reconstruction of the Himalayan orogeny (see Volume 1 – Chapter 1).

1.8. References

Ahmad, T. and Islam, R. (1998). Geochemistry and geodynamic implications of magmatic rocks from the Trans-Himalayan arc. *Geochemical Journal*, 32(6), 383–404.

Aitchison, J.C., Badengzhu Davis, A.M., Liu, J., Luo, H., Malpas, J.G., McDermid, I.R.C., Wu, H., Ziabrev, S.V., Zhou, M.-F. (2000). Remnants of a Cretaceous intra-oceanic subduction system within the Yarlung–Zangbo suture (southern Tibet). *Earth Planetary Science Letters*, 183, 231–244.

Aitchison, J.C., Ali, J.R., Davis, A.M. (2007). When and where did India and Asia collide? *Journal of Geophysical Research*, 112, B05423.

Andjić, G., Zhou, R., Jonell, T.N., Aitchison, J.C. (2022). A single Dras-Kohistan-Ladakh arc revealed by volcaniclastic records. *Geochemistry, Geophysics, Geosystems*, 23, e2021GC010042.

Bard, J.P. (1983). Metamorphism of an obducted island arc; example of the Kohistan Sequence (Pakistan) in the Himalayan collided range. *Earth and Planetary Science Letters*, 65, 133–144.

Baxter, A.T., Aitchison, J.C., Zyabrev, S.V. (2009). Radiolarian age constraints on Mesotethyan ocean evolution, and their implications for development of the Bangong–Nujiang suture, Tibet. *Journal of the Geological Society*, 166(4), 689–694.

Bhat, I.M., Ahmad, T., Rao, D.S. (2019). Origin and evolution of Suru Valley ophiolite peridotite slice along Indus suture zone, Ladakh Himalaya, India: Implications on melt-rock interaction in a subduction-zone environment. *Geochemistry*, 79(1), 78–93.

Borneman, N.L., Hodges, K.V., Van Soest, M.C., Bohon, W., Wartho, J.A., Cronk, S.S., Ahmad, T. (2015). Age and structure of the Shyok suture in the Ladakh region of northwestern India: Implications for slip on the Karakoram fault system. *Tectonics*, 34(10), 2011–2033.

Bosch, D., Garrido, C.J., Bruguier, O., Dhuime, B., Bodinier, J.-L., Padron-Navarta, J.A., Galland, B. (2011). Building an island–arc crustal section: Time constraints from a LA-ICP-MS zircon study. *Earth and Planetary Science Letters*, 309, 268–279.

Bouilhol, P., Burg, J.P., Bodinier, J.L., Schmidt, M.W., Dawood, H., Hussain, S. (2009). Magma and fluid percolation in arc to fore-arc mantle: Evidence from Sapat (Kohistan, Northern Pakistan). *Lithos*, 107, 17–37.

Bouilhol, P., Schaltegger, U., Chiaradia, M., Ovtcharova, M., Stracke, A., Burg, J.-P., Dawood, H. (2011). Timing of juvenile arc crust formation and evolution in the Sapat Complex (Kohistan–Pakistan). *Chemical Geology*, 280, 243–256.

Bouilhol, P., Jagoutz, O., Hanchar, J., Dudas, F. (2013). Dating the India–Eurasia collision through arc magmatic records. *Earth and Planetary Science Letters*, 366, 163–175.

Burg, J.P. (2011). The Asia–Kohistan–India collision: Review and discussion. In *Arc-Continent Collision*, Brown, D. and Ryan, P.D. (eds). Springer, Berlin, Heidelberg.

Burg, J.P. and Bouilhol, P. (2019). Timeline of the South Tibet – Himalayan belt: The geochronological record of subduction, collision, and underthrusting from zircon and monazite U-Pb ages. *Canadian Journal of Earth Sciences*, 56, 1318–1332.

Burg, J.-P., Bodinier, J.L., Chaudrhy, S., Hussain, S., Dawood, H. (1998). Infra-arc mantle–crust transition and intra-arc mantle diapirs in the Kohistan Complex (Pakistani Himalaya); petro-structural evidence. *Terra Nova*, 10, 74–80.

Clift, P.D., Degnan, P.J., Hannigan, R., Blusztajn, J. (2000). Sedimentary and geochemical evolution of the Dras forearc basin, Indus suture, Ladakh Himalaya, India. *Geological Society of America Bulletin*, 112, 450–466.

Corfield, R.I. and Searle, M.P. (2000). Crustal shortening estimates across the north Indian continental margin, Ladakh, NW India. *Geological Society, London, Special Publications*, 170(1), 395–410.

Corfield, R.I., Searle, M.P., Green, O.R. (1999). Photang thrust sheet: An accretionary complex structurally below the Spontang ophiolite constraining timing and tectonic environment of ophiolite obduction, Ladakh Himalaya, NW India. *Journal of the Geological Society*, 156(5), 1031–1044.

Coward, M.P., Rex, D.C., Khan, M.A., Windley, B.F., Broughton, R.D., Luff, I.W., Pudsey, C.J. (1986). Collision tectonics in the NW Himalayas. *Geological Society, London, Special Publications*, 19(1), 203–219.

Debon, F., Le Fort, P., Dautel, D., Sonet, J., Zimmermann, J.L. (1987). Granites of western Karakorum and northern Kohistan (Pakistan): A composite Mid-Cretaceous to upper Cenozoic magmatism. *Lithos*, 20(1), 19–40.

Dhuime, B., Bosch, D., Bodinier, J.L., Garrido, C.J., Bruguier, O., Hussain, S.S., Dawood, H. (2007). Multistage evolution of the Jijal ultramafic–mafic complex (Kohistan, N Pakistan): Implications for building the roots of island arcs. *Earth and Planetary Science Letters*, 261, 179–200.

Dhuime, B., Bosch, D., Garrido, C.J., Bodinier, J.L., Bruguier, O., Hussain, S.S., Dawood, H. (2009). Geochemical architecture of the lower- to middle-crustal section of a paleo-island arc (Kohistan Complex, JijalKamila Area, Northern Pakistan): Implications for the evolution of an oceanic subduction zone. *Journal of Petrology*, 50, 531–569.

Dietrich, V.J., Frank, W., Honegger, K. (1983). A Jurassic-Cretaceous island arc in the Ladakh-Himalayas. *Journal of Volcanology and Geothermal Research*, 18(1–4), 405–433.

Dürr, S.B. (1996). Provenance of Xigaze fore-arc basin clastic rocks (Cretaceous, south Tibet). *Geological Society of America Bulletin*, 108(6), 669–684.

Gaetani, M. (1997). The Karakorum block in central Asia, from Ordovician to Cretaceous. *Sedimentary Geology*, 109(3–4), 339–359.

Gaetani, M., Garzanti, E., Jadoul, F., Nicora, A., Tintori, A., Pasini, M., Khan, K.S.A. (1990). The north Karakorum side of the Central Asia geopuzzle. *Geological Society of America Bulletin*, 102(1), 54–62.

Garrido, C.J., Bodinier, J.L., Dhuime, B., Bosch, D., Chanefo, I., Bruguier, O., Hussain, S.S., Dawood, H., Burg, J.P. (2007). Origin of the island arc Moho transition zone via melt-rock reaction and its implications for intracrustal differentiation of island arcs: Evidence from the Jijal complex (Kohistan complex, northern Pakistan). *Geology*, 35, 683–686.

Garzanti, E. and Van Haver, T. (1988). The Indus clastics: 706 forearc basin sedimentation in the Ladakh Himalaya (India). *Sedimentary Geology*, 59, 237–249.

Green, O.R., Searle, M.P., Corfield, R.I., Corfield, R.M. (2008). Cretaceous-Tertiary carbonate platform evolution and the age of the India-Asia collision along the Ladakh Himalaya (northwest India). *The Journal of Geology*, 116, 331–353.

Grove, T.L., Till, C.B., Krawczynski, M.J. (2012). The role of H2O in subduction zone magmatism. In *Annual Review of Earth and Planetary Sciences*, Jeanloz, R. (ed.). Annual Reviews Inc., Palo Alto.

Hacker, B.R., Kelemen, P.B., Behn, M.D. (2011). Differentiation of the continental crust by relamination. *Earth and Planetary Science Letters*, 307, 501–516.

Hébert, R., Bezard, R., Guilmette, C., Dostal, J., Wang, C.S., Liu, Z.F. (2012). The Indus–Yarlung Zangbo ophiolites from Nanga Parbat to Namche Barwa syntaxes, southern Tibet: First synthesis of petrology, geochemistry, and geochronology with incidences on geodynamic reconstructions of Neo-Tethys. *Gondwana Research*, 22(2), 377–397.

Henderson, A.L., Najman, Y., Parrish, R., BouDagher-Fadel, M., Barford, D., Garzanti, E., Ando, S. (2010). Geology of the Cenozoic Indus Basin sedimentary rocks: Paleoenvironmental interpretation of sedimentation from the western Himalaya during the early phases of India–Eurasia collision. *Tectonics*, 29, TC6015.

Heuberger, S., Schaltegger, U., Burg, J.-P., Villa, I.M., Frank, M., Dawood, H., Hussain, S., Zanchi, A. (2007). Age and isotopic constraints on magmatism along the Karakoram–Kohistan Suture Zone, NW Pakistan: Evidence for subduction and continued convergence after India–Asia collision. *Swiss Journal of Geoscience*, 100, 85–107.

Honegger, K., Dietrich, V., Frank, W., Gansser, A., Thöni, M., Trommsdorff, V. (1982). Magmatism and metamorphism in the Ladakh Himalayas (the Indus-Tsangpo suture zone). *Earth and Planetary Science Letters*, 60(2), 253–292.

Jagoutz, O.E. (2010). Construction of the granitoid crust of an island arc. Part II: A quantitative petrogenetic model. *Contributions to Mineralogy and Petrology*, 160, 359–381.

Jagoutz, O. (2014). Arc crustal differentiation mechanisms. *Earth and Planetary Science Letters*, 396, 267–277.

Jagoutz, O. and Schmidt, M.W. (2012). The formation and bulk composition of modern juvenile continental crust: The Kohistan arc. *Chemical Geology*, 298–299, 79–96.

Jagoutz, O. and Behn, M.D. (2013). Foundering of lower island-arc crust as an explanation for the origin of the continental Moho. *Nature*, 504(7478), 131–134.

Jagoutz, O., Müntener, O., Burg, J.-P., Ulmer, P., Jagoutz, E. (2006). Lower continental crust formation through focused flow in km-scale melt conduits: The zoned ultramafic bodies of the Chilas Complex in the Kohistan Island arc (NW Pakistan). *Earth and Planetary Science Letters*, 242, 320–342.

Jagoutz, O., Müntener, O., Ulmer, P., Pettke, T., Burg, J.-P., Dawood, H., Hussain, S. (2007). Petrology and mineral chemistry of lower crustal intrusions: The Chilas Complex, Kohistan (NW Pakistan). *Journal of Petrology*, 48, 1895–1953.

Jagoutz, O., Burg, J.-P., Hussain S., Dawood, H., Pettke, T., Iizuka, T., Maruyama, S. (2009). Construction of the granitoid crust of an island arc part I: Geochronological and geochemical constraints from the plutonic 1199 Kohistan (NW Pakistan). *Contributions to Mineralogy and Petrology*, 158(6), 739–755.

Jagoutz, O., Müntener, O., Schmidt, M.W., Burg, J.-P. (2011). The respective roles of flux- and decompression melting and their relevant liquid lines of descent for continental crust formation: Evidence from the Kohistan arc. *Earth and Planetary Science Letters*, 303, 25–36.

Jagoutz, O., Bouilhol, P., Schaltegger, U., Müntener, O. (2019). The isotopic evolution of the Kohistan Ladakh arc from subduction initiation to continent arc collision. *Geological Society, London, Special Publications*, 483(1), 165–182.

Jan, M.Q. and Howie, R.A. (1980). Ortho- and clinopyroxenes from the pyroxene granulites of Swat Kohistan, northern Pakistan. *Mineralogical Magazine*, 43, 715–726.

Ji, W.Q., Wu, F.Y., Chung, S.L., Li, J.X., Liu, C.Z. (2009). Zircon U–Pb geochronology and Hf isotopic constraints on petrogenesis of the Gangdese batholith, southern Tibet. *Chemical Geology*, 262(3–4), 229–245.

Kelemen, P.B. (1995). Genesis of High Mg-number andesites and the continental crust. *Contributions to Mineralogy and Petrology*, 120, 1–19.

Khan, M.A., Stern, R.J., Gribble, R.F., Windley, B.F. (1997). Geochemical and isotopic constraints on subduction polarity, magma sources and palaeogeography of the Kohistan Arc, northern Pakistan. *Journal of the Geological Society, London*, 154, 935–946.

Khan, S.D., Walker, D.J., Hall, S.A., Burke, K.C., Shah, M.T., Stockli, L. (2009). Did the Kohistan-Ladakh island arc collide first with India? *Geological Society of America Bulletin*, 121, 366–384.

Kumar, S., Bora, S., Sharma, U.K., Yi, K., Kim, N. (2017). Early Cretaceous subvolcanic calc-alkaline granitoid magmatism in the Nubra-Shyok valley of the Shyok Suture Zone, Ladakh Himalaya, India: Evidence from geochemistry and U–Pb SHRIMP zircon geochronology. *Lithos*, 277, 33–50.

Lakhan, N., Singh, A.K., Singh, B.P., Sen, K., Singh, M.R., Khogenkumar, S., Singhal, S., Oinam, G. (2020). Zircon U-Pb geochronology, mineral and whole-rock geochemistry of the Khardung volcanics, Ladakh Himalaya, India: Implications for Late Cretaceous to Palaeogene continental arc magmatism. *Geological Journal*, 55, 3297–3320.

Mahéo, G., Bertrand, H., Guillot, S., Villa, I.M., Keller, F., Capiez, P. (2004). The south Ladakh ophiolites (NW Himalaya, India): An intraoceanic tholeiitic origin with implication for the closure of the Neo-Tethys. *Chemical Geology*, 203, 273–303.

Mahoney, J.J., Frei, R., Tejada, M.L.G., Mo, X.X., Leat, P.T., Nagler, T.F. (1998). Tracing the Indian Ocean mantle domain through time: Isotopic results from Old West Indian, East Tethyan, and South Pacific seafloor. *Journal of Petrology*, 39, 1285–1306.

Martin, C.R., Jagoutz, O., Upadhyay, R., Royden, L.H., Eddy, M.P., Bailey, E., Weiss, B.P. (2020). Paleocene latitude of the Kohistan–Ladakh arc indicates multistage India–Eurasia collision. *Proceedings of the National Academy of Sciences*, 117(47), 29487–29494.

McDermid, I.R., Aitchison, J.C., Davis, A.M., Harrison, T.M., Grove, M. (2002). The Zedong terrane: A Late Jurassic intra-oceanic magmatic arc within the Yarlung–Tsangpo suture zone, southeastern Tibet. *Chemical Geology*, 187(3–4), 267–277.

Miller, D.J., Loucks, R.R., Ashraf, M. (1991). Platinum-group element mineralization in the Jijal layered ultramafic-mafic complex, Pakistani Himalayas. *Economic Geology*, 86(5), 1093–1102.

Müntener, O. and Ulmer, P. (2018). Arc crust formation and differentiation constrained by experimental petrology. *American Journal of Science*, 318(1), 64–89.

Palin, R.M., Searle, M.P., Waters, D.J., Horstwood, M.S.A., Parrish, R.R. (2012). Combined thermobarometry and geochronology of peraluminous metapelites from the Karakoram metamorphic complex, North Pakistan; new insight into the tectonothermal evolution of the Baltoro and Hunza Valley regions. *Journal of Metamorphic Geology*, 30(8), 793–820.

Parsons, A.J., Hosseini, K., Palin, R.M., Sigloch, K. (2020). Geological, geophysical and plate kinematic constraints for models of the India-Asia collision and the post-Triassic central Tethys oceans. *Earth-Science Reviews*, 208, 103084.

Paterson, S.R. and Ducea, M.N. (2015). Arc magmatic tempos: Gathering the evidence. *Elements*, 11(2), 91–98.

Petterson, M.G. (2019). The plutonic crust of Kohistan and volcanic crust of Kohistan–Ladakh, north Pakistan/India: Lessons learned for deep and shallow arc processes. *Geological Society, London, Special Publications*, 483(1), 123–164.

Petterson, M.G. and Treloar, P.J. (2004). Volcanostratigraphy of arc volcanic sequences in the Kohistan arc, North Pakistan: Volcanism within island arc, back-arc-basin, and intra-continental tectonic settings. *Journal of Volcanology and Geothermal Research*, 130, 147–178.

Petterson, M.G. and Windley, B.F. (1985). Rb-Sr dating of the Kohistan arc-batholith in the trans-Himalaya of north-Pakistan, and tectonic implications. *Earth and Planetary Science Letters*, 74, 45–57.

Phillips, R.J., Parrish, R.R., Searle, M.P. (2004). Age constraints on ductile deformation and long-term slip rates along the Karakoram fault zone, Ladakh. *Earth and Planetary Science Letters*, 226(3–4), 305–319.

Phillips, R.J., Searle, M.P., Parrish, R.R. (2013). The geochemical and temporal evolution of the continental lithosphere and its relationship to continental-scale faulting: The Karakoram Fault, eastern Karakoram, NW Himalayas. *Geochemistry, Geophysics, Geosystems*, 14(3), 583–603.

Plank, T. and Langmuir, C.H. (1988). An evaluation of the global variations in the major element chemsitry of arc basalts. *Earth and Planetary Science Letters*, 90, 349–370.

Plank, T. and Langmuir, C.H. (1998). The chemical composition of subducting sediment and its consequences for the crust and mantle. *Chemical Geology*, 145, 325–394.

Pudsey, C.J. (1986). The Northern Suture, Pakistan: Margin of a Cretaceous island arc. *Geological Magazine*, 123(4), 405–423.

Ravikant, V., Wu, F.Y., Ji, W.Q. (2009). Zircon U–Pb and Hf isotopic constraints on petrogenesis of the Cretaceous–Tertiary granites in eastern Karakoram and Ladakh, India. *Lithos*, 110(1–4), 153–166.

Raz, U. and Honegger, K. (1989). Magmatic and tectonic evolution of the Ladakh block from field studies. *Tectonophysics*, 161(1–2), 107–118.

Reuber, I. (1989). The Dras arc: Two successive volcanic events on eroded oceanic crust. *Tectonophysics*, 161(1–2), 93–106.

Reuber, I., Colchen, M., Mevel, C. (1987). The geodynamic evolution of the south-Tethyan margin in Zanskar, NW Himalaya, as revealed by the Spontang ophiolitic mélange. *Geodinamica Acta*, 1, 283–296.

Reubi, O. and Müntener, O. (2022). Making andesites and the continental crust: Mind the step when wet. *Journal of Petrology*, in press.

Rex, A.J., Searle, M.P., Tirrul, R., Crawford, M.B., Prior, D.J., Rex, D.C., Barnicoat, A. (1988). The geochemical and tectonic evolution of the central Karakoram, north Pakistan. *Philosophical Transactions of the Royal Society of London. Series A, Mathematical and Physical Sciences*, 326(1589), 229–255.

Ringwood, A.E. (1974). The petrological evolution of island arc systems. *Journal of the Geological Society*, 130, 183–204.

Robertson, A.H. and Collins, A.S. (2002). Shyok Suture Zone, N Pakistan: Late Mesozoic–Tertiary evolution of a critical suture separating the oceanic Ladakh Arc from the Asian continental margin. *Journal of Asian Earth Sciences*, 20(3), 309–351.

Robinson, A.C. (2009). Geologic offsets across the northern Karakorum fault: Implications for its role and terrane correlations in the western Himalayan-Tibetan orogen. *Earth and Planetary Science Letters*, 279(1–2), 123–130.

Rolland, Y. (2002). From intra-oceanic convergence to post-collisional evolution: Example of the India-Asia convergence in NW Himalaya, from Cretaceous to present. *Journal of the Virtual Explorer*, 8, 193–216.

Rolland, Y., Pêcher, A., Picard, C. (2000). Middle Cretaceous back-arc formation and arc evolution along the Asian margin: The Shyok Suture Zone in northern Ladakh (NW Himalaya). *Tectonophysics*, 325(1–2), 145–173.

Rolland, Y., Mahéo, G., Guillot, S., Pêcher, A. (2001). Tectono-metamorphic evolution of the Karakorum Metamorphic complex (Dassu–Askole area, NE Pakistan): Exhumation of mid-crustal HT–MP gneisses in a convergent context. *Journal of Metamorphic Geology*, 19(6), 717–737.

Rolland, Y., Picard, C., Pêcher, A., Carrio, E., Sheppard, S.M., Oddone, M., Villa, I.M. (2002). Presence and geodynamic significance of Cambro-Ordovician series of SE Karakoram (N Pakistan). *Geodinamica Acta*, 15(1), 1–21.

Rolland, Y., Villa, I.M., Guillot, S., Mahéo, G., Pêcher, A. (2006a). Evidence for pre-Cretaceous history and partial Neogene (19–9 Ma) reequilibration in the Karakorum (NW Himalayan Syntaxis) from 40Ar–39Ar amphibole dating. *Journal of Asian Earth Sciences*, 27(4), 371–391.

Rolland, Y., Carrio-Schaffhauser, E., Sheppard, S.M.F., Pêcher, A., Esclauze, L. (2006b). Metamorphic zoning and geodynamic evolution of an inverted crustal section (Karakorum margin, N Pakistan), evidence for two metamorphic events. *International Journal of Earth Sciences*, 95(2), 288–305.

Rolland, Y., Mahéo, G., Pêcher, A., Villa, I.M. (2009). Syn-kinematic emplacement of the Pangong metamorphic and magmatic complex along the Karakorum Fault (N Ladakh). *Journal of Asian Earth Sciences*, 34(1), 10–25.

Rudnick, R.L. (1995). Making continental-crust. *Nature*, 378, 571–578.

Rudnick, R.L. and Gao, S. (2003). Composition of the continental crust. In *Treatise on Geochemistry*, Rudnick, R.L. (ed.). Elsevier, Amsterdam.

Schaltegger, U., Zeilinger, G., Frank, M., Burg, J.P. (2002). Multiple mantle sources during island arc magmatism; U–Pb and Hf isotopic evidence from the Kohistan arc complex, Pakistan. *Terra Nova*, 14, 461–468.

Schärer, U., Xu, R.H., Allègre, C.J. (1984). UPb geochronology of Gangdese (Transhimalaya) plutonism in the Lhasa-Xigaze region, Tibet. *Earth and Planetary Science Letters*, 69(2), 311–320.

Searle, M.P. and Hacker, B.R. (2019). Structural and metamorphic evolution of the Karakoram and Pamir following India–Kohistan–Asia collision. *Geological Society, London, Special Publications*, 483(1), 555–582.

Searle, M.P., Rex, A.J., Tirrul, R., Rex, D.C., Barnicoat, A., Windley, B.F. (1989). Metamorphic, magmatic, and tectonic evolution of the central Karakoram in the Biafo-Baltoro-Hushe regions of northern Pakistan. *Geological Society of America Special Papers*, 232. doi: 10.1130/SPE232.

Searle, M.P., Parrish, R.R., Tirrul, R., Rex, D.C. (1990). Age of crystallization and cooling of the K2 gneiss in the Baltoro Karakoram. *Journal of the Geological Society*, 147, 603–606.

Singh, S., Kumar, R., Barley, M.E., Jain, A.K. (2007). SHRIMP U-Pb ages and depth of emplacement of Ladakh Batholith, eastern Ladakh, India. *Journal of Asian Earth Sciences*, 30, 490–503.

St-Onge, M.R., Rayner, N., Searle, M.P. (2010). Zircon age determinations for the Ladakh batholith at Chumathang (Northwest India): Implications for the age of the India–Asia collision in the Ladakh Himalaya. *Tectonophysics*, 495(3–4), 171–183.

Tahirkheli, R.A.K. (1979). Geology of Kohistan and adjoining Eurasia and Indio-Pakistan continents, Pakistan. *Geological Bulletin, University of Peshawar*, 11, 1–30.

Taylor, S.R. and White, A.J.R. (1965). Geochemistry of andesites and growth of continents. *Nature*, 208, 271.

Treloar, P.J., Rex, D.C., Guise, P.G., Coward, M.P., Searle, M.P., Windley, B.F., Luff, I.W. (1989). K-Ar and Ar–Ar geochronology of the Himalayan collision in NW Pakistan: Constraints on the timing of suturing, deformation, metamorphism and uplift. *Tectonics*, 8(4), 881–909.

Ulmer, P., Kaegi, R., Muntener, O. (2018). Experimentally derived intermediate to silica-rich arc magmas by fractional and equilibrium crystallization at 1.0 GPa: An evaluation of phase relationships, compositions, liquid lines of descent and oxygen fugacity. *Journal of Petrology*, 59, 11–58.

Upadhyay, R., Frisch, W., Siebel, W. (2008). Tectonic implications of new U–Pb zircon ages of the Ladakh Batholith, Indus suture zone, northwest Himalaya, India. *Terra Nova*, 20, 309–317.

Villiger, S., Ulmer, P., Muntener, O., Thompson, A.B. (2004). The liquid line of descent of anhydrous, mantle-derived, tholeiitic liquids by fractional and equilibrium crystallization – An experimental study at 1 center dot 0 GPa. *Journal of Petrology*, 45, 2369–2388.

Walsh, J.M.J., Buckman, S., Nutman, A.P., Zhou, R.J. (2019). Age and Provenance of the Nindam Formation, Ladakh, NW Himalaya: Evolution of the Intraoceanic Dras Arc before collision with India. *Tectonics*, 38, 3070–3096.

Walsh, J.M., Buckman, S., Nutman, A.P., Zhou, R. (2021). The significance of Upper Jurassic felsic volcanic rocks within the incipient, intraoceanic Dras Arc, Ladakh, NW Himalaya. *Gondwana Research*, 90, 199–219.

Weinberg, R.F. and Dunlap, W.J. (2000). Growth and deformation of the Ladakh Batholith, Northwest Himalayas: Implications for timing of continental collision and origin of calc-alkaline batholiths. *The Journal of Geology*, 108(3), 303–320.

Weinberg, R.F., Dunlap, W.J., Whitehouse, M. (2000). New field, structural and geochronological data from the Shyok and Nubra valleys, northern Ladakh: Linking Kohistan to Tibet. *Geological Society, London, Special Publications*, 170(1), 253–275.

White, L.T., Ahmad, T., Ireland, T.R., Lister, G.S., Forster, M.A. (2011). Deconvolving episodic age spectra from zircons of the Ladakh Batholith, northwest Indian Himalaya. *Chemical Geology*, 289(3–4), 179–196.

Yang, Y.T., Guo, Z.X., Luo, Y.J. (2017). Middle-Late Jurassic tectonostratigraphic evolution of Central Asia, implications for the collision of the Karakoram-Lhasa Block with Asia. *Earth-Science Reviews*, 166, 83–110.

Yoshino, T. and Okudaira, T. (2004). Crustal growth by magmatic accretion constrained by metamorphic P-T paths and thermal models of the Kohistan arc, NW Himalayas. *Journal of Petrology*, 45, 2287–2302.

Zhu, D.C., Mo, X.X., Niu, Y., Zhao, Z.D., Wang, L.Q., Pan, G.T., Wu, F.Y. (2009). Zircon U–Pb dating and in-situ Hf isotopic analysis of Permian peraluminous granite in the Lhasa terrane, southern Tibet: Implications for Permian collisional orogeny and paleogeography. *Tectonophysics*, 469(1–4), 48–60.

2
Suture Zone

Julia DE SIGOYER[1] and Jean-Luc EPARD[2]
[1] University Grenoble Alpes, France
[2] University of Lausanne, Switzerland

2.1. Introduction

What is the suture zone in the Himalaya? The suture zone is a term borrowed from medicine vocabulary. It describes the structure joining tissues or bones that were previously separated and have fused together. In geology, the suture zone represents the zone between two former continents after they have merged together during the collision to form a mountain belt. Argand et al. (1924) was the first to understand that the suture zone represents the past oceanic domain. The recognition of the suture zone was an important step to understand the plate tectonics theory.

The suture zone represents therefore the internal part of a belt. It is generally strongly deformed and highly heterogeneous as it contains structures and elements of various origins. The suture zone builds up as an interpenetrative stacking of ophiolitic slices being the relics of the past oceanic lithosphere intermixed with many other units, such as volcanic island (arc volcanism or OIB), sedimentary units, accretionary, wedge and continental deposits from episutural basin. All these units are the witness of a part of the oceanic history from its birth to its closure. The heterogeneity and great diversity of this zone reflect then the complexity of the events that occurred before the collision from rift drift through accretionary subduction to collision in between the two continents.

In the Himalaya, several suture zones are described (Figure 2.1):

– The main one is the Indus Yarlung Tsangpo (or Zangbo) suture zone (ITSZ) that runs along the 2,500 km of the internal part of the Himalayan belt from Pakistan to Namche Barwa (Figure 2.1). It actually represents the boundary in between the Asian plate to the north and the Indian plate to the south, which used to be the ocean Neotethys and its sub-basins. The width of the ITSZ can vary from a few hundred meters to several tens of kilometers (Figure 2.1). ITSZ name derives from the Indus river that wanders along the suture zone.

– Another type of suture, less continuous, comes from the fact that oceanic lithosphere has locally thrust towards the south on top of the northern Indian passive margin, on the Permian to Paleocene–Eocene sediments (Tethyan Himalayan unit). Several obducted ophiolites are then recognized along the Himalayan arc. The Spongtang ophiolite klippe in Ladakh (Figure 2.2), the Kiogar–Yungbwa ophiolite in SW Tibet and some minor ones the Xiugugabu, Dangxiong, Zhongba ophiolites.

– North of the Ladakh batholith lies the Shyok suture zone that separates the Karakorum batholith intruded on the Asian continental margin from the Kohistan–Ladakh batholith that was a magmatic arc developed either on oceanic or continental crust. The Shyok suture is then considered as a back arc ophiolite (see Volume 2 – Chapter 1 for more information).

The occurrence of several suture zones (ITSZ and Shyok) attests for several subduction zones that allowed the closure of the oceanic domain in between Asia and India.

Only the Indus Tsangpo suture zone will be described in detail in this chapter.

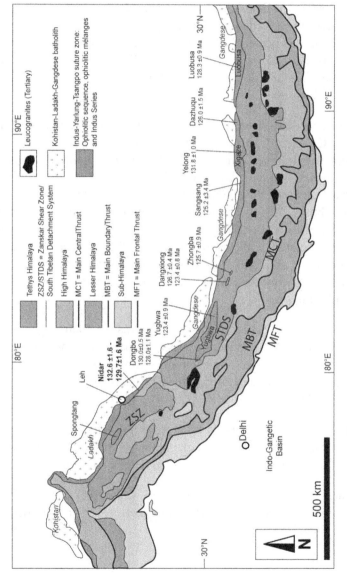

Figure 2.1. Simplified tectonic map of the Himalaya (Buchs 2018) modified from Gansser (1964). U/Pb zircon ages from ophiolites are shown on this map; see Buchs (2018) for references. For a color version of this figure, see www.iste.co.uk/cattin/himalaya2.zip

2.2. General geological description of the ITSZ

All along the ITSZ, occurrences of Neotethyan oceanic lithosphere are observed together with deep-marine sediments, fore-arc sediments, seamonts, mélange zone, accretionary wedge and Indus episutural sediments emplaced during the closure of the oceanic domain with provenance from different paleogeography. The suture zone lies between the Tethyan sediments of the Indian margin to the south and the Karakorum–Ladakh–Lhasa batholith to the north.

The Neotethys ocean has opened from Late Carboniferous to Late Permian (see Volume 2 – Chapter 3 for more details on the rifting phase). The closure of the Neotethys initiated during Lower Cretaceous as an intra-oceanic north-dipping subduction (Mahéo et al. 2004; Hébert et al. 2012; Buchs 2018). The ophiolites scattered along the ITSZ are distributed into two groups of ages: the Spongtang, Luobusa, Zedang and Kiogar ophiolites have Jurassic–Lower Cretaceous ages based on radiochronologic dating (Catlos et al. 2019) and age of sediments associated with the ophiolite; all the other are of Lower Cretaceous age (Hébert et al. 2012), based on the Sm/Nd and U/Pb zircon ages of the dyke of diabase or gabbros (Figure 2.1), and by the ages of radiolarite that overlay the ophiolite sequences.

Ophiolites along the ITSZ can differ in terms of petrological and geochemical aspects but all of them present intra-oceanic suprasubduction characteristics such as the fore-arc and back-arc settings. Only the Spongtang ophiolite seems to present also a previous phase of magmatism typical for the MORB signature (Catlos et al. 2019).

Some of the ITSZ ophiolites present a complete section of the "Penrose-type" oceanic lithosphere when other ones are dismembered ophiolites that present only fractions (often the mantle part) of the sequence.

Most ophiolites were formed in short-lived (30 Myr) basins. Many reconstructions propose a magmatic arc associated with an intraoceanic subduction close to the Eurasian continent.

2.3. The Indus suture zone of the Eastern Ladakh, the Nidar zone

One of the largest and best preserve outcrops of the ITSZ is in the Nidar area located in eastern Ladakh at about 150 km SE from Leh (Figure 2.2). Two valleys crosscut the suture zone Mahe–Sumdo and from Nidar–Kyun valleys, they correspond to the two cross-sections represented in Figure 2.3.

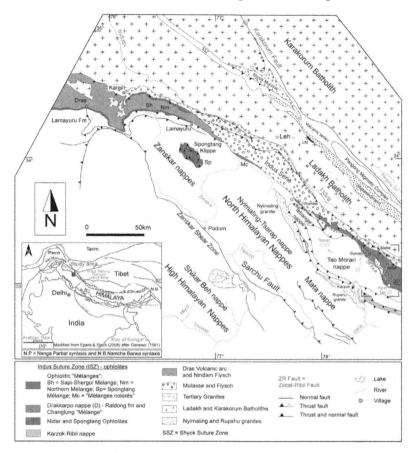

Figure 2.2. *Tectonic map of the NW Himalaya modified after Thakur and Virdi (1979), Thakur and Misra (1984), Robertson and Degnan (1994), Steck et al. (1998), Robertson (2000), Rolland et al. (2000), Rolland et al. (2002), Steck (2003), Epard and Steck (2008) and Buchs and Epard (2019). The Nidar ophiolite is represented by the open red rectangle on the tectonic map. For a color version of this figure, see www.iste.co.uk/cattin/himalaya2.zip*

In eastern Ladakh, the ITSZ is composed from north to south: (1) the Indus sequence, (2) the Nidar ophiolite and (3) the accretionary wedge (or mélange zone) containing the Drakkarpo and Ribil units. The ITSZ is locally separated from the Indian margin by the Zildat normal fault (de Sigoyer et al. 2004; Buchs 2018).

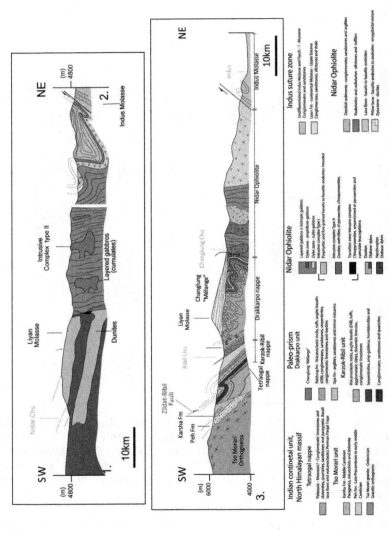

Figure 2.3. *Two cross-sections across the ITSZ in Nidar area. Sections 1 and 2 represent the situation for the right (SE) side of the Nidar Chu; section 3 is located 15 km NW of 1 and 2. Modified from Buchs and Epard (2019). For a color version of this figure, see www.iste.co.uk/cattin/himalaya2.zip*

2.3.1. *The Indus group sediments*

The Indus sedimentary sequence composed of continental shale, conglomerates and sandstones was deposited in an episutural basin, the intramontane Indus basin that evolved from a marine to a continental environment and recorded the closure of the marine domain between the two converging continents.

In the Nidar area, the ophiolite as well as other tectonic units of the suture zone are unconformably overlain by Late Eocene continental deposits (Liyan Formation). This places an upper age limit for major tectonic movements in the Indus suture zone.

The sequence lies on the Ladakh batholith that was still active 60 Ma ago. The age of the beginning of the sequence is discussed between late Cretaceous to Paleocene; it finishes with Miocene conglomerates.

2.3.2. *The Nidar ophiolite*

The Nidar ophiolite can be studied along the Mahe–Sumdo section (Figure 2.3 section 3); the ophiolitic sequence is highly reduced in comparison with the Nidar–Kyun Tso section (Figures 2.2 and 2.3 section 1.2). The quality of the outcrops and the low deformation of the rocks have allowed detailed mapping (Buchs and Epard 2019) of this unit. It has not been subducted at depth, nor buried under other tectonic units during the Himalayan orogeny and has therefore escaped from the Himalayan metamorphism. It is only metamorphosed under greenschist condition that could correspond to the hydration of the oceanic lithosphere during an oceanic context. According to its chemical composition, the Nidar ophiolite is interpreted as a suprasubduction ophiolite (de Sigoyer 1998; Mahéo et al. 2004; Ahmad et al. 2008; Buchs 2018).

The formations will be described from north to south enabling a complete description of the oceanic lithosphere from top to bottom (Figures 2.2–2.4):

– Various detrital sedimentary rocks are observed on the top of the ophiolite. Barremian–Aptian radiolarites lay just on top of the pillow lava. Conglomerates overlay the radiolarites, with a higher proportion of volcanic arc pebbles at the top of the serie, and a higher proportion of pebbles coming from the Nidar ophiolite at the base of the sedimentary serie.

– Pillow lavas are the most abundant volcanic rocks and are located in between the sediment and the layered gabbros. The pillow lava zone is locally intruded by dacitic metrics dykes.

– The gabbro unit is about 5 km thick, with layered gabbros at the base. The layers are formed by the variable abundance of pyroxene. Towards the top, the gabbros become isotropic and are also more altered. The isotropic gabbros in the upper part have similar mineralogy as the layered gabbros (layers of anorthosites and gabbronorites containing plagioclases, pyroxene and olivine), but in general, they are more altered and veined. The layered and isotropic gabbros are cut by two types of intrusive complexes (type I and type II). The first is formed by amphibole gabbros and the second (type II) by partially serpentinized ultramafics (dunites and clinopyroxenite), gabbros associated with metric pockets of plagiogranites (see Figure 2.3, 2.4). The plagiogranites have been dated at 131 Ma (Hauterivian) by U–Pb on zircons (see below). According to their geochemical signatures, the intrusions are related to an arc emplacement installed on the suprasubduction ophiolite (Buchs 2018).

– Under the gabbros lie ultramafic rocks from the mantle. Partially serpentinized (2–60%) harzburgites are observed with channels of dunites in the surrounding harzburgites. The serpentinization is higher close to the contact with the gabbros. Harzburgites have equigranular textures composed of orthopyroxene (20–40%), olivine (30–60%), spinel (2–5%) and rare clinopyroxenes (<3%). Dunites have coarse-grained textures made of large olivine (up to 3 cm) and small amounts of orthopyroxenes (<2%). The field relations allow us to reconstitute the following history for the ultramafic unit: (a) formation of harzburgites locally replaced by dunites and (b) the dunites are in turn cut by chromites and then by clinopyroxenites and small gabbro dykes. Diabase dykes represent the most recent magmatic activity (Figure 2.3).

– The field observations indicate that the Nidar ophiolite underwent at least two magmatic events: (i) phase 1 magmatism regroups a classical ophiolitic sequence with residual harzburgites, layered and isotropic gabbros, pillow lavas and the lava flows of the volcanosedimentary unit and the dykes in pillow lavas; and phase 2 magmatism regroups dunites, chromites and clinopyroxenites in the ultramafic unit, the intrusive complexes type I and II, the dykes in the upper part of the gabbro unit and diabases in the upper part of the ultramafic unit.

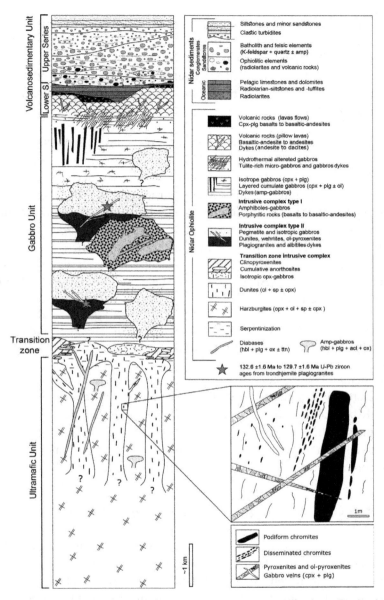

Figure 2.4. *Lithostratigraphic profile across the Nidar ophiolite from Buchs (2018). The 1 km scale is for the ophiolite and not for the overlying sediments. For a color version of this figure, see www.iste.co.uk/cattin/himalaya2.zip*

Bulk rock geochemical data from ophiolitic sequence and intrusive complexes of the Nidar ophiolite (Figure 2.5) display LILE enrichment relative to REE and HFSE, Nb-Ta negative anomalies, low Ti/V and relatively high Th/Nb ratios. Such geochemical signatures are widely observed in fore-arc setting and attests that the Nidar ophiolite represents the magmatic lithosphere in a suprasubduction zone context.

Zircon U–Pb dates of four plagiogranites that intruded the Nidar ophiolite show ages from 132.6 ±1.6 Ma to 129.7 ±1.6 Ma (Figure 2.6). Isochrones on four-layered Nidar gabbros resulted in a less constrained Sm/Nd model age of 140 ± 32 Ma (Ahmad et al. 2008). Considering that the layered gabbros belong to phase I of the formation of oceanic crust; its formation is comprised between 140 and 132 Ma. Amphibole Ar–Ar ages (120 - 110 Ma, Mahéo et al. 2004) from the upper part of the gabbro unit could date a stage of hydrothermal metamorphism. These findings suggest that the Nidar ophiolite records the magmatic build-up of a new intraoceanic island arc crust related to the Lower Cretaceous subduction of oceanic crust.

The Nidar ophiolite was first thrust to the south over the Drakkarpo unit and then backthrusted towards the north over the Indus Serie.

Continental coarse conglomerates are observed in between the Nidar ophiolite and the Drakkarpo unit. They contain pebbles of both units, as well as from Ribil and Tso Morari units (from the south). This formation contains the products of erosion of the local unit after the collision. Similar deposits are observed in the Western Ladakh Shergol Conglomerates, and also in South Tibet where they were dated at Oligocene time (Van Haver 1984).

2.3.3. *The accretionary wedge or oceanic mélange*

Accretionary wedge or prism is formed above the subduction trench by the accretion of sedimentary and magmatic rocks from the upper oceanic lithosphere. The accretionary wedge prism can also contain metamorphic rocks exhumated from the subducting zone (see Volume 2 – Chapter 4). South of the Nidar ophiolite, two units of mélange were identified: the Drakkarpo unit and the Ribil–Karzog unit (de Sigoyer et al. 2004; Buchs and Epard 2019).

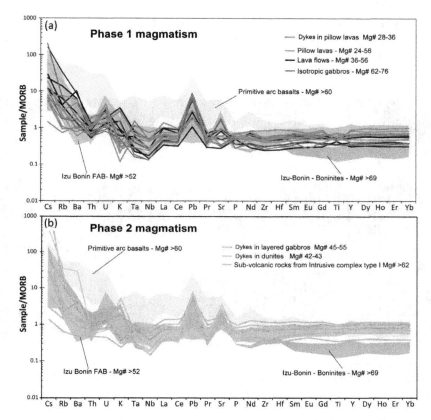

Figure 2.5. Whole rock trace elements diagrams normalized to MORB, modified from Buchs (2018). Phase 1 magmatism = oceanic crust formation and phase 2 magmatism = arc-related intrusions. (a) Nidar ophiolite pillow lavas, lava flows and isotropic gabbros of phase 1 magmatism and porphyritic sub-volcanic rocks (intrusive complexes type I) of the phase 2 magmatism. (b) Dykes in layered gabbros, dykes in pillow lavas and diabases in dunites of phase 2 magmatism. Primitive arc basalt field is from Kelemen et al. (2003), fore-arc basalt and boninite fields are from Reagan et al. (2010). For a color version of this figure, see www.iste.co.uk/cattin/himalaya2.zip

2.3.3.1. The Drakkarpo unit

The Drakkarpo unit was first recognized as the Zildat ophiolitic mélange, Thakur and Virdi (1979) and then interpretated as a part of an accretionary wedge by de Sigoyer (1998) and Buchs (2018). The Drakkarpo unit is discontinuously outcropping along 5 km wide and 150 km long area from the Kyun Tso region to the Rumste village (Figures 2.2 and 2.3). Its base consists of several hundreds of meters of shales and sandstones (Gya Formation) followed by a thick polygenic conglomerates (Raldong Formation, Buchs 2018). Its matrix is composed of clay level, green sandstones or calcareous slates in which lenses of tuffs, augite-OIB basalts, serpentinites, quartzites, micaschists and radiolarites are observed.

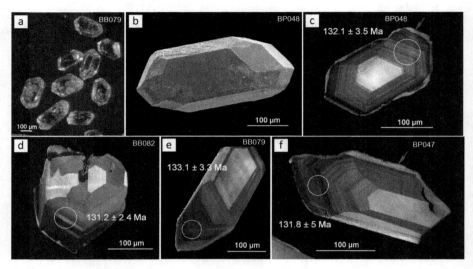

Figure 2.6. *Dating zircons from Buchs (2018). (a) Inclusion-free zircons of the Nidar ophiolite. (b) SEM image Au-covered Zircon. (c–f) Cathodoluminescence images (CL) of polished zircons. White circles on each picture are the locations of LA-ICPMS analysis showing $^{206}Pb/^{238}U$ ages. Spot sizes are 35 μm for BP045 and 50 μm for BP048, BB079 and BB082. Red arrows indicate small overgrowth rims. For a color version of this figure, see www.iste.co.uk/cattin/himalaya2.zip*

The holes and fractures of the tuffs are filled with carbonates, chlorites and oxides suggesting hydrothermal metamorphism. These volcanic rocks have alkaline affinities; their geochemical features on the volcanic rocks suggest an oceanic island (OIB) origin (Fuchs and Linner 1997; de Sigoyer 1998). Thick white limestones typical of platform facies environment are observed; their ages and paleogeographical location are still uncertain. These limestones are embedded in Upper Albian to Mid-Cenomanien red sandstones (Fuchs and Linner 1996).

A mélange, the "Changlung mélange" is defined in the northern part of the Drakkarpo nappe (Buchs 2018). It is made up of blocks and small clasts of volcanosedimentary rocks, pillow lavas, radiolarites, serpentinites, breccias, red marls and limestones eroded from the underlying part of Drakkarpo nappe.

The matrix of the "mélange" is formed by sandstones and contains blocks of Upper Ypresian nummulite-bearing limestones (Buchs 2018). Pebbles of volcanic rocks from the overlying Nidar ophiolite and Ladakh Batholith are present in conglomerates of the "Changlung Mélange".

The location of the Drakkarpo unit, between the Nidar ophiolite and the Indian margin, the succession of the different lithologies, and the geochemical data suggest that the Drakkarpo unit represents a remnant of a former Upper Cretaceous seamount such as the Photang unit observed below the Spontang ophiolite (Reuber 1990), accreted in the accretionary wedge.

The Tertiary Changlung mélange was deposited during upper Ypresian in a piggy-back basin on the developing accretionary wedge.

2.3.3.2. *The Karzog and Ribil units*

The Karzok–Ribil nappe is composed of the Karzog ophiolites (Berthelsen 1953) and the Ribil unit (de Sigoyer 1998). Buchs (2018) considers that the Karzok–Ribil is a nappe that has thrusted towards the south above the Tso Morari crystalline dome (see Volume 2 – Chapter 4). The Karzok–Ribil nappe is composed of slices of an ophiolitic sequence (serpentinized peridotites, gabbros and pillow lavas), chromite, radiolarites, marbles and dolomitic marbles, polygenic conglomerates, breccia and agglomeratic slates from the Indian margin. The Karzog unit mostly consists of chromitic pods that lies in between the Mata unit and the Tso Morari Ultra High pressure unit (see Volume 2 – Chapter 4). The Ribil unit really looks like the Drakkarpo unit. It contains augite-basalts (OIB) and volcanosedimentary rocks. This unit is metamorphosed under greenschist conditions. Brachiopod fragments from the Upper Paleozoic (Upper Carboniferous to Permian) were found in the marbles by Fuchs and Linner (1996). A similar succession of rocks is described in the basement of the Lamayuru formation (Colchen et al. 1994) and that represents the distal part of the Indian continental margin. Therefore, the Ribil unit is interpreted as being originally derived from an oceanic island arc, a seamount, located close to the Indian passive margin. This unit was underthrust below the Drakkarpo unit in the accretionary wedge.

2.3.3.3. *The Zildat Shear Zone*

The ITSZ is separated from the eclogitic Tso Morari unit that belongs to the Indian margin by the Zildat shear zone along which 100 m thick serpentinite layers are formed. The occurrence of metamorphic olivine and talc in serpentinites suggests that the serpentinization took place at about (600°C, 20 kbar) in the serpentinized channel that lies above the subduction zone (Guillot et al. 2000). Such serpentinites form part of the ITZS. These serpentinites are considered by de Sigoyer (1998), Guillot et al. (2000) and de Sigoyer et al. (2004) as being part of the subduction channel that contribute to the exhumation of the UHP Tso Morari unit, when Steck et al. (1998) and (Buchs and Epard 2019) consider that they are part of the Ribil–Karzog unit cross-cut by the Zildat brittle fault.

2.4. Conclusion

The studies on the suture zones in the Himalaya bring essential data on the geodynamic evolution during the formation and closure or the oceanic domain that separated the Indian from the Asian continent. Studies of the age and provenance of the sediments that lies on the episutural basins attested for the closure of marine domain at the upper Eocene time follow by the Indian–Asian collision and continental sedimentation in the episutural basin.

The ITSZ in eastern Ladakh (with the Nidar ophiolite) presented in this chapter is very representative of most of the ITSZ all along the Himalayan belt. It represents part of the Neotethyan oceanic lithosphere formed during Lower Cretaceous in an intraoceanic subduction that generate a magmatic arc and a back arc oceanic accretion. The Nidar ophiolite has recorded two stages of magmatism in this supra subduction context. The first stage generates a fore-arc oceanic crust when the second stage was represented by intrusive rocks in a back arc that intruded the formed oceanic lithosphere (Buchs 2018). The subduction-related magmatism migrated to the north between Lower Cretaceous to Mid-Eocene forming a volcanic arc. There is no consensus yet whether this arc is connected to Dras arc to the Ladakh batholith or not (see Volume 2 – Chapter 1). The Nidar suture zone also contains an accretionary wedge formed during the subduction processes. Seamounts such as the Drakkarpo and Ribil units were accreted on this accretionary wedge too. While HP rocks were described in other ITSZ (see Volume 2 – Chapter 4 for details), only one occurrence of glaucophane was described in the accretionary wedge of the Nidar area by (Virdi 1986) but it has never been confirmed. The accretionary wedge was formed before late Ypresian as attested by the age of Changlung mélange that overlaid the accretion wedge.

At the scale of the Himalayan belt, we note that some slices of Neotethyan oceanic lithosphere were obducted onto the northern margin of India. The age of this obduction is still discussed and could be either at Latest Cretaceous–Early Tertiary 65–62 Ma (Ding et al. 2005) in South Central Tibet or at about 50 Ma in Ladakh (Clift et al. 2014) based on the dating of the sediment that are below the Spongtang ophiolite. The southward obduction of ophiolite seems to mark the demise of the Neotethys intraoceanic subduction system followed by the initial India–Asia contact during late Paleocene.

All studies conducted along the ITSZ are consistent in proposing at least one intraoceanic subduction of the Neotethys towards the North, south of the Kohistan–Ladakh–Gangdese batholith. Most of the ophiolites observed along the ITSZ were formed in this context and present Cretaceous ages (132–123 Ma). The initiation of subduction is in between 150 and 134 Ma (Jagoutz et al. 2019); the Nidar ophiolite reflects similar range of ages.

Many studies propose two intra-oceanic subduction zones south of the Ladakh Gangdese magmatic arc, to form the batholith, the Dras arc, the suprasubduction ophiolite such as the Nidar one, and the obducted ophiolite such as the Songpang klippe in Ladakh. At least one of this intraoceanic subduction was located in the northern part of the ocean close to the Ladakh arc (see Figure 2.7 for a geodynamic reconstruction based on the Nidar section (Buchs 2018)).

Another subduction is required towards the north below the Asian active margin (Karakorum to the west and Lhasa block in the central part of the belt) to explain the Shyok suture zone (see Volumes 1 – Chapter 1 and Volume 2 – Chapter 1 for more information on different types of geodynamic reconstructions).

The observation of several suture zones and several magmatic arcs in the internal part of the Himalaya requires that multiple intraoceanic subduction zones were activated simultaneously or alternatively since the Cretaceous to close the Neotethys ocean. Most of the ophiolites observed in the Himalaya are related to these subduction processes. The complex subduction system requires to close the Neotethys looks like the present subduction of the Pacific, Mariana and Philippine oceanic plates.

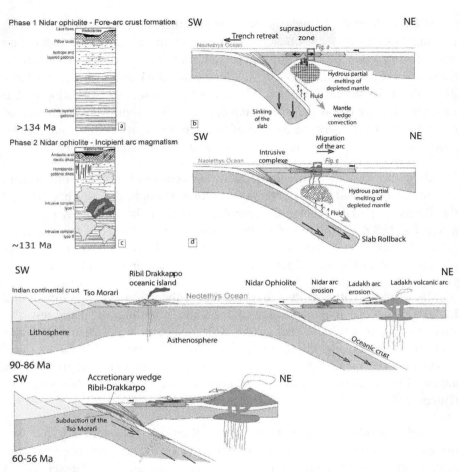

Figure 2.7. *Schematic geodynamic reconstruction of the Neotethys subduction based on the observations done on the Nidar ITSZ from lower Cretaceous (134 Ma) to Paleocene (60–56 Ma) modified from Buchs (2018). a and c represent the lithostratigraphic profiles of the oceanic lithosphere formed during phases 1 and 2, respectively. b and c represent the geodynamic reconstruction during phases 1 and 2 of the formation of the Nidar oceanic lithosphere. For a color version of this figure, see www.iste.co.uk/cattin/himalaya2.zip*

2.5. References

Ahmad, T., Tanaka, T., Sachan, H.K., Asahara, Y., Islam, R., Khanna P.P. (2008). Geochemical and isotopic constraints on the age and origin of the Nidar ophiolitic complex, Ladakh, India. Implications for the Neo-Tethyan subduction along the indus suture zone. *Tectonophysics*, 451(1–4), 206–224.

Argand, E. (1924). La tectonique de l'Asie. *Compte-rendu du XIIIe congrès géologique international*, 171–372.

Berthelsen, A. (1953). On the Geology of the Rupshu District, N.W. Himalaya. Medd. Dansk Geol., Foren, Kobenhavn, 12, 351–414.

Buchs, N. (2018). Geology of the Nidar – Tso Morari area (Indian Himalayas, Ladakh): From intra-oceanic subduction to nappe exhumation. PhD Thesis, University of Lausanne.

Buchs, N. and Epard, J.-L. (2019). Geology of the eastern part of the Tso Morari nappe, the Nidar Ophiolite and the surrounding tectonic units, NW, Himalaya, India. *Journal of Maps*, 15(2), 38–48.

Catlos, E.J., Pease, E.C., Dygert, N., Brookfield, M., Schwarz, W.H., Bhutani, R., Pande, K., Schmitt, A. (2019). Nature, age, and emplacement of the Spongtang Ophiolite, Ladakh, NW India. *Journal of the Geological Society*, 176(2), 284–305.

Clift, P.D., Carter, A., Jonell, T.N. (2014). U–Pb dating of detrital zircon grains in the Paleocene stumpata formation, Tethyan Himalaya, Zanskar, India. *Journal of Asian Earth Sciences*, 82, 80–89.

Colchen, M., Mascle, G., Delaygue, G. (1994). Lithostratigraphy and age of the formations in the Tso Morari dome (abstract). *Journal of the Geological Society of Nepal*, 10, 23.

Ding, L., Kapp, P., Wan, X. (2005). Paleocene–Eocene record of ophiolite obduction and initial India–Asia collision, south central Tibet. *Tectonics*, 24(3). doi:10.1029/2004TC001729.

Epard, J.-L. and Steck, A. (2008). Structural development of the Tso Morari ultra-high pressure nappe of the Ladakh Himalaya. *Tectonophysics*, 451(1–4), 242–264.

Fuchs, G. and Linner, M. (1996). On the geology of the suture zone and Tso Morari Dome in Eastern Ladakh (Himalaya). *Jahrbuch der Geologischen Bundesanstalt Sonderband*, 139(2), 191–207.

Fuchs, G. and Linner, M. (1997). Multiphase tectonics in the Indus Suture Zone of Eastern Ladakh [Abstract]. *12th Himalayan Karakorum Tibet Workshop*, 33–35.

Gansser, A. (1964). *Geology of the Himalayas*. John Wiley, London.

Guillot, S., Hattori, K., de Sigoyer, J. (2000). Mantle wedge serpentinization and exhumation of eclogites: Insights from eastern Ladakh, northwest Himalaya. *Geology*, 28(3), 199–202.

Hébert, R., Bezard, R., Guilmette, C., Dostal, J., Wang, C.J., Liu, Z.F. (2012). The Indus–Yarlung Zangbo ophiolites from Nanga Parbat to Namche Barwa Syntaxes, Southern Tibet: First synthesis of petrology, geochemistry, and geochronology with incidences on geodynamic reconstructions of Neo-Tethys. *Gondwana Research*, 22(2), 377–397.

Jagoutz, O., Bouilhol, P., Schaltegger, U., Müntener, O. (2019). The isotopic evolution of the Kohistan Ladakh arc from subduction initiation to continent arc collision. In *Himalayan Tectonics: A Modern Synthesis*, Treloar, P.J. and Searle, M.P. (eds). Geological Society, Special Publications, London.

Kelemen, P.B., Hanghøj, K., Greene, A.R. (2003). One view of the geochemistry of subduction-related magmatic arcs, with an emphasis on primitive andesite and lower crust. *Treatise on Geochemistry*, 3, 659.

Mahéo, G., Bertrand, H., Guillot, S., Villa, I.V., Keller, F., Capiez, P. (2004). The South Ladakh ophiolites (NW Himalaya, India): An intra-oceanic tholeiitic arc origin with implication for the closure of the Neo-Tethys. *Chemical Geology*, 203(3–4), 273–303.

Reagan, M.K., Ishizuka, O., Stern, R.J., Kelley, K.A., Ohara, Y., Blichert-Toft, J., Bloomer, S.H., Cash, J., Fryer, P., Hanan, B.B. et al. (2010). Fore-arc basalts and subduction initiation in the Izu-Bonin-Mariana System. *Geochemistry, Geophysics, Geosystems*, 11(3), doi.org/10.1029/2009GC002871.

Reuber, I. (1990). Géométrie et dynamique de l'accrétion dans les ophiolites téthysiennes : Himalaya du ladakh, Oman et Turquie. Thesis, Université de Bretagne occidentale.

Robertson, A. (2000). Formation of mélanges in the Indus Suture Zone, Ladakh Himalaya by successive subduction-related, collisional and post-collisional processes during Late Mesozoic-Late Tertiary time. *Geological Society, London, Special Publications*, 170(1), 333–374.

Robertson, A. and Degnan, P. (1994). The Dras Arc complex: Lithofacies and reconstruction of a Late Cretaceous oceanic volcanic arc in the Indus Suture Zone, Ladakh Himalaya. *Sedimentary Geology*, 92(1), 117–145.

Rolland, Y., Pêcher, A., Picard, C. (2000). Middle Cretaceous back-arc formation and arc evolution along the Asian margin: The Shyok Suture Zone in northern Ladakh (NW Himalaya). *Tectonophysics*, 325(1), 145–173.

Rolland, Y., Picard, C., Pêcher, A., Lapierre, H., Bosch, D., Keller, F. (2002). The cretaceous Ladakh arc of NW Himalaya – Slab melting and melt–mantle interaction during fast northward drift of Indian Plate. *Chemical Geology*, 182(2), 139–178.

de Sigoyer, J. (1998). Mécanismes d'exhumation des roches de haute pression basse température en contexte de convergence continentale (Tso Morari, NO Himalaya). Doctoral Dissertation, Université Claude Bernard-Lyon I.

de Sigoyer, J., Guillot, S., Dick, P. (2004). Exhumation of the ultrahigh-pressure Tso Morari unit in Eastern Ladakh (NW Himalaya): A case study. *Tectonics*, 23(3), doi.org/10.1029/2002TC001492.

Steck, A. (2003). Geology of the NW Indian Himalaya. *Eclogae Geologicae Helvetiae*, 96(2), 147–196.

Steck, A., Epard, J.-L., Vannay, J.-C., Hunziker, J., Girard, M., Morard, A., Robyr, M. (1998). Geological transect across the Tso Morari and Spiti Areas: The Nappe structures of the Tethys Himalaya. *Eclogae Geologicae Helvetiae*, 91(1), 103–122.

Thakur, V.C. and Misra, D.K. (1984). Tectonic framework of the Indus and Shyok suture zones in Eastern Ladakh, Northwest Himalaya. *Tectonophysics*, 101(3–4), 207–220.

Thakur, V.C. and Virdi, N.S. (1979). Lithostratigraphy, structural framework, deformation and metamorphism of the southeastern region of Ladakh, Kashmir Himalaya, India. *Himalayan Geology*, 9, 63–78.

Van Haver, T. (1984). Etude stratigraphique sédimentologique et structurale d'un bassin d'avant arc : exemple du Bassin de l'Indus, Ladakh, Himalaya. PhD Thesis, Univ. Joseph Fourier, Grenoble.

Virdi, N.S. (1986). Indus-Tsangpo suture in the Himalaya-crustal expression of a Palaeo-Subduction Zone. *Annale Societatis Geologorum Poloia*, 56, 3–31.

3
Geological Evolution of the Tethys Himalaya

Chiara MONTOMOLI[1], Jean-Luc EPARD[2], Eduardo GARZANTI[3], Rodolfo CAROSI[1] and Martin ROBYR[2]

[1] *University of Turin, Italy*
[2] *University of Lausanne, Switzerland*
[3] *Università degli Studi di Milano-Bicocca, Milan, Italy*

3.1. Introduction

The Tethys Himalaya (TH), also known as "Tibetan Zone", is one of the major tectonic domains within the Himalayan orogen, stretching for approximately 2,000 km from Kashmir and the Zanskar–Spiti synclinorium to South Tibet. Its northern boundary coincides with the Indus–Tsangpo Suture Zone (see Volume 2 – Chapter 2), separating the Tethys Himalaya from the Transhimalayan arc-trench system and Lhasa Block to the north (Gansser 1980), whereas the southern boundary is a tectonic contact with the Greater Himalaya Sequence (GHS), commonly referred to as South Tibetan Detachment System (STDS) (Burg et al. 1984; Herren 1987, Figure 3.1, see Volume 1 – Chapter 3).

The sedimentary succession, one of the most complete and spectacularly exposed on Earth, represents the deformed remnants of the northern continental margin of India and preserves an over 500 Ma-long stratigraphic record (Gaetani and Garzanti 1991; Garzanti 1999). The succession can be subdivided into a pre-Neotethyan (Neoproterozoic to earliest Mississippian) and a Neotethyan part. The latter includes a rift stage (Mississippian to lowermost Permian), documenting the tectonic and magmatic processes that culminated in continental break-up and ocean-floor spreading, followed by a drift stage (upper Lower Permian to earliest Paleocene) recording the subsidence history of the northern passive continental margin of Gondwana. Identified within the drift stage are an initial *starved passive-margin phase* (upper Lower Permian to Middle Triassic), a *Late Triassic extensional event*, a *Jurassic mature passive-margin stage* and a *Cretaceous volcanic and drowning event*. During the final *collision stage* (Paleocene–Eocene), the passive margin of India collided with the active margin of Eurasia (Sciunnach and Garzanti 2012) (Figure 3.2). The Tethys Himalaya zone is much larger in South Tibet, where it is traditionally divided into a southern part characterized by coastal to shelfal sedimentation and an outer part consisting of offshore and deep-water deposits (Hu et al. 2008).

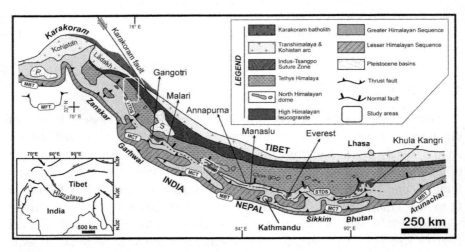

Figure 3.1. *Schematic map of the Himalayan belt and location of the study areas. For a color version of this figure, see www.iste.co.uk/cattin/himalaya2.zip*

During the construction of the orogen, the Tethys Himalaya underwent a polyphase deformation history characterized by the development of several systems of folds and related thrusts and ductile – brittle shear zones.

The Tethys Himalaya experienced, generally, very-low to low-grade metamorphic conditions during deformation. Higher metamorphic conditions have been detected for the lower portion of the Tethys Himalaya, which discontinuously crops out along the belt reaching amphibolite facies conditions in the south (Garzanti and Brignoli 1989; Garzanti et al. 1992; Spring and Crespo-Blanc 1992; Schneider and Masch 1993; Dunkl et al. 2011; Montomoli et al. 2017).

In this chapter, after a general description of the stratigraphy of the Tethys Himalaya, we will focus on two different sections, located in Ladakh (NW India) and Dolpo (Western Nepal) where the sequence is spectacularly exposed (Figure 3.1).

3.2. The stratigraphy of the Tethys Himalaya

3.2.1. *The pre-Tethyan history*

The upper Neoproterozoic succession of the NW Himalaya testifies to passive-margin sedimentation. Fine-grained tidal-flat siltstones and sandstones pass upward to a peritidal carbonate platform, followed in turn by shales containing Middle Cambrian trilobites and thin-bedded deeper-water sandstones. A similar succession is exposed at the top of Mount Everest, 1,130 km to the SE (Myrow et al. 2009). An angular unconformity overlain by thick alluvial-fan conglomerates testifies to a major Cambro-Ordovician tectonic event, related to the Pan-African orogeny documented over large parts of Gondwana (Garzanti et al. 1986).

In Nepal and south Tibet, the Ordovician is instead documented by very thick, sparsely fossiliferous shallow-water carbonates followed by quartzarenites and calcschists (Colchen et al. 1986). The Silurian and Devonian are represented by a shallow-marine mixed carbonate-siliciclastic unit overlain by the coastal Muth Quartzarenite in the NW Himalaya (Draganits and Noffke 2004). Graptolite-bearing mudrocks followed by tentaculitid-bearing marlstones and carbonates capped by an ironstone interval overlain by gray mudrocks with intercalated sandstones are exposed in the Nepal Himalaya (Colchen et al. 1986; Figures 3, 4 and 10 in Garzanti et al. (1992)). Widespread fossiliferous limestones deposited in the early Mississippian are truncated by a major disconformity testifying to the onset of Neotethyan rifting (Figure 11 in Garzanti et al. 1992).

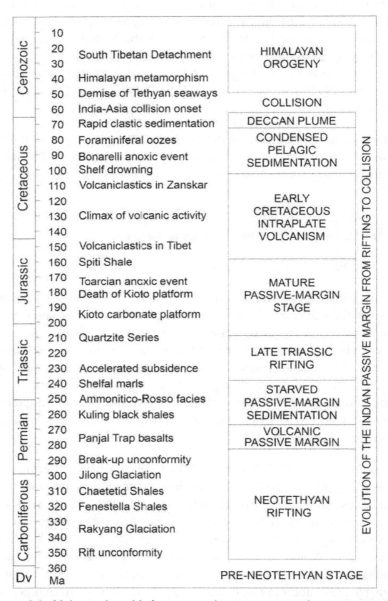

Figure 3.2. *Main stratigraphic features and tectono-magmatic events recorded in the Tethys Himalaya succession (from Sciunnach and Garzanti (2012) and other sources cited in the text)*

3.2.2. *The Neotethyan rift stage*

The Mississippian to lowermost Permian rift sequence displays marked lateral changes in thickness and facies indicating sedimentation in tectonically controlled basins (Garzanti et al. 1994c, 1996a) and includes dikes of bimodal alkalic rocks and volcanic detritus (Vannay and Spring 1993; Caironi et al. 1996; Sciunnach and Garzanti 1997).

In the NW Himalaya, the syn-rift succession begins with a gypsum horizon containing middle Tournaisian/early Visean brachiopods, suggesting deposition in half-grabens with restricted circulation in arid settings (Figure 5 in Gaetani et al. 1986; Draganits et al. 2002). The overlying dark mudrocks with intercalated quartzose sandstones are capped by glacigenic diamictites.

In South Tibet, four diamictite-bearing intervals containing faceted pebbles, trapezoidal cobbles and boulders $\leq$ 1,000 m^3 in volume indicate that shore ice began to grow as early as the Visean–Serpukhovian, triggered by basin inversion and tectonic uplift during initial rifting at middle southern latitudes (Garzanti and Sciunnach 1997).

After the Mississippian glacial stage, two transgressive events are documented by black-shale marker units of early Pennsylvanian (Fenestella Shales; Hayden 1994) and middle Pennsylvanian age (Chaetetid Shales; Garzanti et al. 1998a). A renewed, more extensive glaciation is testified by glacio-marine diamictites deposited from Kashmir to South Tibet during the Asselian climax of rifting. Rift-shoulder uplift resulted in the extensive erosion of the Gondwanan margin and development of disconformities corresponding to major gaps (Sciunnach and Garzanti 1996) or even cutting into pre-rift strata. The entire rift sequence is thus missing locally (e.g. Fuchs 1977).

3.2.3. *The Neotethyan drift stage*

The base of the drift sequence is represented either by a major disconformity with an erosion surface reaching down to the Devonian, Silurian or even Ordovician or, in thicker sections, by a paraconformity mantled by condensed horizons of hybrid and ferruginous arenites dated biostratigraphically as mid-Sakmarian (Gaetani et al. 1990; Garzanti et al. 1996b; Sakagami et al. 2006). These transgressive sediments reflect the combined effect of thermal subsidence of the newly formed rifted margin and sea-level rise fostered by deglaciation.

The lower part of the drift sequence includes continental flood basalts exposed discontinuously along the newly formed volcanic passive margin from Kashmir to South Tibet (Garzanti et al. 1999). In other localities, the succession includes quartz-rich sandstones passing upward to brachiopod-rich black mudrocks. Strongly condensed biocalcarenites rich in brachiopods and conodonts were deposited in offshore settings in western Nepal to south Tibet at the close of the Permian (Nicora and Garzanti 1997).

3.2.3.1. *Starved passive-margin sedimentation*

Condensed pelagic deposition became widespread in the Lower Triassic. Three intervals of ammonoid-bearing nodular limestone are either separated by black shales intervals or amalgamated. The total thickness ranges from 33–54 m in the NW Himalaya to 6.5–20 m in South Tibet, corresponding to low accumulation rates (1–10 m/Ma (Garzanti et al. 1994b, 1995, 1998b). The overlying ammonoid-bearing marls testify to the increasing terrigenous supply and consequently increased accumulation rates from $\leq$ 10 m/Ma in the Middle Triassic to $\sim$20 m/Ma in the Carnian.

3.2.3.2. *Late Triassic rifting*

Sedimentation rates increased notably in the Norian (50–100 m/Ma). Shallow-water carbonates with intercalated sandstones and oolitic ironstones were deposited in the NW Himalaya, whereas thick mudrocks with phosphatic nodules and interbedded, locally bioclastic feldspatho-quartzose sandstones accumulated in Nepal to South Tibet (Garzanti et al. 1995; Jadoul et al. 1998). The occurrence of felsic and mafic volcanic rock fragments with alkalic signature, together with zircon grains yielding U–Pb ages of 229–223 Ma, indicates penecontemporaneous magmatic activity in an intraplate extensional setting. Coeval block-faulting and volcanism in northwestern Australia (von Rad et al. 1992) suggest that a rift-related volcanic belt developed along the northern margin of eastern Gondwana, controlling the accumulation of thick turbidities of the Langjiexue Group in the central-eastern Himalaya and of the Lamayuru Group in the NW Himalaya (Wang et al. 2016; Meng et al. 2021).

The close of the Triassic is marked by the widespread deposition of quartz-rich coastal sandstones with intercalated oo-bioclastic arenites, chamosite-goethite ironstones and mudrocks, and initial establishment of a carbonate platform documented by Megalodon-bearing limestones in the NW Himalaya (Jadoul et al. 1990).

3.2.3.3. The Jurassic mature passive-margin stage

The Lower Jurassic is characterized by the widespread deposition of the Kioto platform carbonates, beginning later in the east than in the west and characterized by Lithiothis-bearing horizons (Jadoul et al. 1998). Accumulation rates progressively decreased to less than 10 m/Ma in the earliest Middle Jurassic, reflecting reduced thermal subsidence and outbuilding of the Gondwanan continental terrace during the mature passive margin stage.

A major paleogeographic change was recorded around the Bajocian/Bathonian boundary, when the drowning of the Kioto carbonate platform was followed by the deposition of hybrid arenites, marls and "lumachelle" layers on a storm-dominated shelf (Laptal Formation; Heim and Gansser 1939). Drastic variations in thickness and megaslumps documented by gigantic olistoliths of Kioto carbonates found in the continental-rise succession (Bassoullet et al. 1981) indicate a major block-faulting event, possibly related to the rifting of India–Madagascar from Africa in the west and from NW Australia in the east, followed by sea floor spreading in the Somali, Mozambique and Argo basins (Patriat et al. 1982; von Rad et al. 1992).

The disconformity at the top of the Kioto Group may be directly onlapped by transgressive horizons of the late Bathonian to the middle Callovian Ferruginous Oolite Formation, which represents a widespread, condensed oolitic ironstone horizon traced all along the Tethys Himalaya (Garzanti 1993a). The Upper Jurassic is documented by offshore, ammonoid-rich black mudrocks, representing a long established and most characteristic stratigraphic interval (Spiti Shale; Stoliczka 1866). In large parts of Nepal, the Spiti Shale is the youngest stratigraphic unit exposed at the core of tight synclines.

3.2.3.4. Early Cretaceous volcanism and mid-Cretaceous drowning

A drastic change in sedimentation from offshore shales to locally coal-bearing deltaic and shelfal siliciclastic rocks took place close to the Jurassic/Cretaceous boundary. In the NW Himalaya, a lower unit characterized by feldspatho-quartzose sandstones capped by a glauconitic horizon is followed by volcaniclastic sandstones interbedded in proximal areas with up to very coarse-grained feldspatho-quartzose sandstones (Giumal Group; Garzanti 1992, 1993b). A few ammonoid-bearing intervals document the late Berriasian/early Valanginian and late Aptian/early Albian (Bertle and Suttner 2021). In the Thakkhola graben of Nepal, a basal quartzarenite unit is disconformably followed by coarse-grained deltaic volcaniclastic deposits containing wood logs, overlain in turn by dark shelfal volcaniclastic mudrocks and locally glauconitic sandstones yielding ammonoids of Aptian age (Chukh Group; Bordet et al. 1971; Garzanti and Pagni Frette 1991; Gibling et al. 1994).

In South Tibet, radiolarian-rich phosphatic volcanic arenites are overlain by coarsening-upward sequences of locally bioclastic sandstones showing hummocky lamination or intensely burrowed, alternating with intervals of dark gray/greenish siltstones containing carbonate concretions (Wölong Volcaniclastics; Jadoul et al. 1998; Hu et al. 2010).

Volcanic detritus reached South Tibet (Tithonian) earlier than Nepal (Valanginian) and much earlier than the Spiti–Zanskar synclinorium (Aptian/Albian). The U–Pb ages of zircon grains found in Lower Cretaceous units range from $\geq$ 110 to $\leq$ 150 Ma and cluster at $\sim$130 Ma (Hu et al. 2015a), covering the periods of magmatic activity in the Comei province of southeastern Tibet and in the Rajmahal province of eastern India (Baksi et al. 1987; Zhu et al. 2009; Singh et al. 2020). The intraplate character of volcanic source rocks is documented by dominant mafic detritus including pebbles with alkalic geochemical signature in older strata (Dürr and Gibling 1994), associated up-section with microlitic to felsitic and microgranitoid rock fragments suggesting both the changing character of magmatic activity and the incipient unroofing of subvolcanic bodies (Garzanti 1999; Hu et al. 2010). The still poorly understood westward delay in the onset of volcaniclastic sedimentation testified by the stratigraphic record was explained with either progressive westward migration of extensional/transtensional tectonic and magmatic activity or gradual progradation of sediment fed from eastern volcanic centers (Hu et al. 2015a). These tectono-magmatic events marked the final break-up of Gondwana and opening of the Indian Ocean.

Volcaniclastic sedimentation ended synchronously in the late Albian (Premoli Silva et al. 1991). Drowning of the clastic shelf is testified by a condensed section spanning the latest Albian to Cenomanian and represented by glauconitic arenites and channelized coarse-grained glauco-phosphorite intervals up to 35-m-thick in the inner margin (Garzanti et al. 1989). The overlying gray to multicolored, Turonian to Campanian pelagic limestones are capped by a major hiatus ascribed to the initial impingement of the Deccan plume-head at the base of the Indian lithosphere (Garzanti and Hu 2015). A thick regressive sequence of bioclastic marls and quartz rich sandstones documents a major pulse of terrigenous supply and drastic increase in accumulation rates during the Maastrichtian (Willems et al. 1996; Hu et al. 2012).

3.2.4. The Paleocene–Eocene collision stage

The Paleocene begins with the deposition of coastal quartzarenites capped by condensed glauconitic layers and overlain by thick shallow-marine carbonates with rich benthic foraminiferal faunas spanning the Selandian and Thanetian (Nicora et al. 1987; Li et al. 2015). A major disconformity locally associated with intrabasinal carbonate conglomerates and straddling the Paleocene/Eocene boundary is documented both in the NW Himalaya and South Tibet. Long interpreted as the far-field effect of collision onset and ascribed to uplift caused by a flexural wave propagating from the suture zone (Garzanti et al. 1987), the unconformity may also reflect the climate change during the Paleocene/Eocene thermal maximum (Li et al. 2017).

In the Eocene, drowning of the carbonate platform was unconformably followed by deltaic green and red beds fed from Transhimalayan volcanic and ultramafic rocks, thus documenting the final closure of Neotethyan seaways (Najman et al. 2010, 2017). Arrival of the distal edge of the Tethys Himalayan margin at the Transhimalayan trench and consequent initiation of continental subduction beneath ophiolitic forearc crust has been recently constrained by radiometric dating of tuff layers, biostratigraphy and detrital-zircon chronostratigraphy as between 59 and 61 Ma (DeCelles et al. 2014; Hu et al. 2015b; An et al. 2021).

3.3. Deformation of the Tethys Himalaya

Deformation of the Tethys Himalaya will be described in two different key areas of the belt, Dolpo (Western Nepal) and Ladakh (NW India), where the sequence offers very good outcrops and continuous sections (Figure 3.1).

3.3.1. Deformation and metamorphism of the Tethys Himalaya in Dolpo (Western Nepal)

In this area (Figure 3.3), the Tethys Himalaya has been affected by different tectonic events giving rise to several systems of folds developing at different scales.

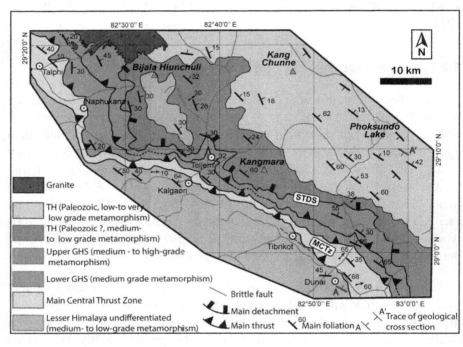

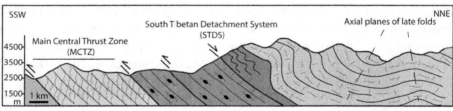

Figure 3.3. *Geological map of the Dolpo area (Western Nepal) and related geological cross-section. For a color version of this figure, see www.iste.co.uk/cattin/himalaya2.zip*

In particular, it is possible to recognize three different tectonic events, named D_1, D_2 and D_3. In the lower part of the sequence, composed of metalimestone, calcschist and quartzite of the Dhaulagiri Formation (Frank and Fuchs 1970; Fuchs 1977), all the three deformation phases can be recognized.

The first D_1 tectonic event, related to continental collision and underthrusting of the Greater Himalayan Sequence below the Tethys Himalaya, is characterized by open to tight metric to centimetric folds, deforming the bedding surface. D_1 folds have an S–SW vergence. Folds are associated with the development of an axial plane foliation (S_1). S_1 foliation is much more well developed in pelitic levels, where it is a fine continuous foliation, whereas in the more competent levels, it is a spaced rough, sometimes anastomosing cleavage. S_1 is characterized by the dynamic recrystallization of isoriented minerals such as muscovite, biotite, chlorite, quartz, calcite, and oxides.

The second D_2 tectonic event is associated with centimetric to hectometric asymmetric folds showing NE vergence (Figure 3.4). These folds have been recognized mainly approaching the STDS, contrary to other sections of the belt where D_2 folds are regionally developed and are related to a still ongoing compressive tectonics (Montomoli et al. 2017).

A_2 axes show variable orientations but they mostly trend NW–SE dipping a very few degrees towards the NW. An S_2 foliation develops parallel to the axial plane of folds and it ranges from a continuous foliation to a crenulation cleavage. Approaching the lower South Tibetan Detachment System, the S_2 foliation becomes much better defined, and strikes parallel to the shear zone attitude.

The last D_3 tectonic event is characterized by the occurrence of hectometric to kilometric open upright folds affecting all tectonic units, as well as the STDS. These late folds have axes trending NW–SE and a foliation parallel to their axial planes, with no dynamic recrystallization.

In this area, the STDS is characterized only by the lower ductile shear zone (Carosi et al. 2002) and no evidence of the upper brittle fault, recognized in other sections (Carosi et al. 1998; Searle 1999) has been recognized.

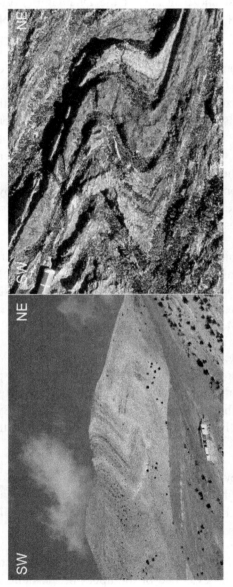

Figure 3.4. *Example of the D2 asymmetric fold developed at different scales. For a color version of this figure, see www.iste.co.uk/cattin/himalaya2.zip*

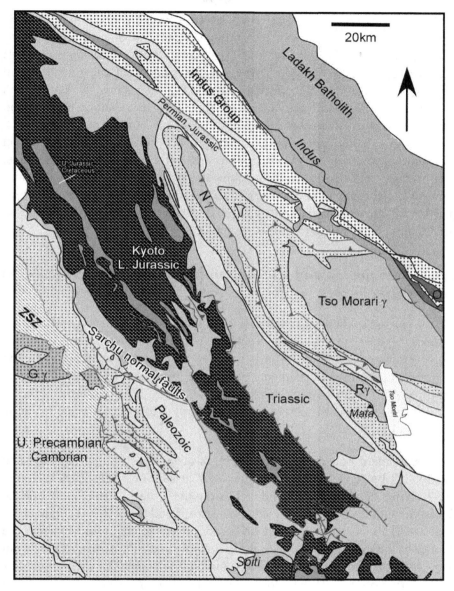

Figure 3.5. *Geologic map of the Ladakh area between Sarchu and the Indus Valley (modified from Steck 2003). ZSZ: Zanskar Shear Zone; Rγ Rupshu granite; Nγ: Nyimaling granite. For a color version of this figure, see www.iste.co.uk/cattin2/himalaya2.zip*

In the Dolpo area, the TH has been deformed under very low to low-grade metamorphic conditions estimated through illite and chlorite crystallinity (Garzanti et al. 1994a; Crouzet et al. 2007) and vitrinite reflectance methods (Crouzet et al. 2007). By the way, in the study section, moving from the upper part of the TH down to the STDS, we observe an increase in the metamorphic grade and a change from low-temperature deformation mechanisms to crystalline plasticity both in quartz and calcite (Carosi et al. 2002, 2007). In the upper part of the TH, pressure solution deformation mechanisms predominate, affecting fossil shell valves as well as detrital grains. Approaching the marbles at the bottom of the sequence, foliation becomes pervasive, and quartz and calcite grains accommodate deformation by crystal plasticity. Calcite deformation twins also indicate a downward increase in deformation temperature (Schill et al. 2002).

3.3.2. Deformation and metamorphism of the Tethys Himalaya in Ladakh (NW India)

The section studied is located in NW India, in the state (Union Territory) of Ladakh (Figure 3.1). On this transect, the Tethys Himalaya is limited by two major structures. To the north, it is bounded by the Indus suture, comprising tectonic units formed by elements of oceanic origin and sealed by Cenozoic deposits of the Indus Group (Henderson et al. 2010). This boundary is almost vertical and locally corresponds to a normal fault or to a dextral strike slip zone (Epard and Steck 2008).

The southern boundary is marked by structures considered equivalent to the South Tibetan Detachment System: the Zanskar Shear Zone (ZSZ : Herren 1987; Dèzes et al. 1999). This ductile structure is cut by high-angle normal faults known locally as the Sarchu faults (Epard and Steck 2004) (Figures 3.5 and 3.6). This normal faults and shear zones (top to the N movements) are related to the emplacement of tectonic units forming the Greater Himalayan Sequence and dome formation (see Volume 2 – Chapter 6). These structures are superimposed to thrust movements related to the emplacement of the nappe stack forming the Tethys Himalaya. In the absence of a well-marked shear zone, the boundary between the Tethys Himalaya and the GHS becomes diffuse and difficult to locate precisely. This is particularly the case in the vicinity of Sarchu and to the SE of that locality (Figure 3.5).

Geological Evolution of the Tethys Himalaya 69

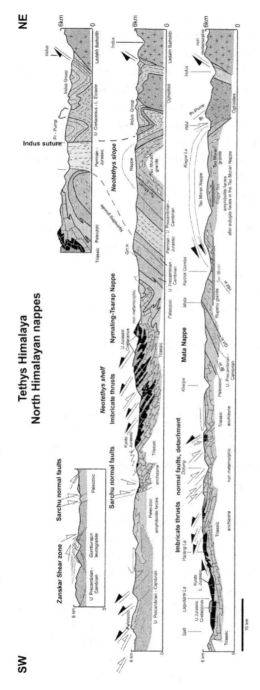

Figure 3.6. *Cross-sections of the Tethys Himalaya in Ladakh (modified from Steck 2003; Epard and Steck 2004). For a color version of this figure, see www.iste.co.uk/cattin/himalaya2.zip*

The tectonic structures of the Tethys Himalaya are formed at the expense of the northern part of the Indian margin. They are grouped under the name of North Himalayan nappes (Figure 3.6). The Nyimaling–Tsarap–Mata nappe is formed by sediments of the Neotethyan shelf and their Precambrian–Paleozoic substratum (Figure 3.6). Imbricate thrusts with Mesozoic carbonate rocks are characteristic of the southern, frontal part, whereas towards the North, sediments of the Neotethyan slope are involved in the North Himalayan nappe stack (Steck et al. 1998). One of them is the Tso Morari nappe extensively studied due to its UHP metamorphism. It represents the leading edge of the thinned Indian plate that has been subducted approximately 55 Ma (high-pressure mineral assemblage, de Sigoyer et al. (2000)). The Tso Morari nappe is exposed in the core of a dome structure formed by backfolds and late "en échelon" structures, the latter related to dextral strike-slip shear zone parallel to the Indus Suture (Epard and Steck 2008).

The Barrovian metamorphism associated with the North Himalayan nappes emplacement ranges from very low-grade in the hanging-wall of the Zanskar Shear Zone to amphibolite facies (kyanite–staurolite zone with secondary sillimanite) in the Tso Morari dome. This metamorphic overprint has been dated at 48-45 Ma by de Sigoyer et al. (2000) (see Volume 2 – Chapter 4).

3.4. Conclusion

Spectacularly exposed in the Tethys Himalaya is one of the most complete sedimentary successions on Earth, documenting the complete evolution of the Indian passive margin of Neotethys from rifting in the Carboniferous to collision in the Paleocene. The Tethys Himalaya forms a domain that can be followed along the entire Himalayan chain. It is characterized by a series of S-verging nappes. Its Northern boundary with the Indus–Tsangpo ophiolitic suture is locally marked by normal faults that elevate the southern compartment and allow the exhumation of the deepest and most metamorphic parts of the nappe stack. This is the case, for example, of the Tso Morari tectonic unit. Towards the south, the frontal thrusts of the nappes are reactivated and cut by the South Tibetan Detachment System and its lateral equivalents such as the Zanskar Shear Zone in Ladakh. This is a ductile shear zone that lowers the northern compartment and favors the preservation from erosion of the highest and least metamorphic tectonic units of the Tethys Himalaya. The STDS that marks the boundary between the Tethys Himalaya and Greater Himalaya Sequence is not a continuous localized shear zone and presents a complex architecture made by ductile and brittle ductile shear zones. Locally, the normal fault-like movement is distributed over a series of steeply dipping faults. The boundary between TH and GHS is therefore no longer a single structure, but a zone several tens of kilometers wide. The precise position of the STDS or its lateral equivalents is sometimes a matter of controversy.

3.5. References

An, W., Hu, X., Garzanti, E., Wang, J.G., Liu, Q. (2021). New precise dating of the India–Asia collision in the Tibetan Himalaya at 61 Ma. *Geophysical Research Letters*, 48(3), e2020GL090641.

Baksi, A.K., Barman, T.R., Paul, D.K., Farrar, E. (1987). Widespread early Cretaceous flood basalt volcanism in eastern India: Geochemical data from the Rajmahal-Bengal-Sylhet traps. *Chemical Geology*, 63(1–2), 133–141.

Bassoullet, J.P., Colchen, M., Marcoux, J., Mascle, G. (1981). Les masses calcaires du flysch Triasico-Jurassique de Lamayuru (Zone de la suture de l'Indus, Himalaya du Ladakh) : klippes sédimentaires et élements de plate-forme remaniés. *Rivista Italiana di Paleontologia e Stratigrafia*, 86(4), 825–844.

Bertle, R.J. and Suttner, T.J. (2021). Facies, sequence stratigraphy and depositional environment of the Lower Cretaceous Giumal Formation in Spiti (Tethyan Himalaya, India). *Marine and Petroleum Geology*, 128, 105009.

Bordet, P., Colchen, M., Krummenacher, D., Le Fort, P., Mouterde, R., Remy, M. (1971). *Recherches géologiques dans l'Himalaya du Népal, région de la Thakkhola*. C.N.R.S., Paris.

Burg, J.P., Brunel, M., Gapais, D., Chen, G.M., Liu, G.H. (1984). Deformation of leucogranites of the crystalline Main Central Sheet in southern Tibet (China). *Journal of Structural Geology*, 6, 535–542.

Caironi, V., Garzanti, E., Sciunnach, D. (1996). Typology of detrital zircon as a key to unravelling provenance in rift siliciclastic sequences (Permo-Carboniferous of Spiti, N India). *Geodinamica Acta*, 9(2), 101–113.

Carosi, R., Lombardo, B., Molli G., Musumeci, G., Pertusati, P.C. (1998). The South Tibetan detachment system in the Rongbuk valley, Everest region. Deformation features and geological implications. *Journal of Asian Earth Sciences*, 16, 299–331.

Carosi, R., Montomoli, C., Visonà, D. (2002). Is there any detachment in the Lower Dolpo (western Nepal)? *Comptes rendus géoscience*, 334(12), 933–940.

Carosi, R., Montomoli, C., Visonà, D. (2007). A structural transect in the Lower Dolpo: Insights on the tectonic evolution of Western Nepal. *Journal of Asian Earth Sciences*, 29(2–3), 407–423.

Colchen, M., Le Fort, P., Pêcher, A. (1986). *Annapurna, Manaslu, Ganesh Himal*. C.N.R.S., Paris.

Crouzet, C., Dunkl, I., Paudel, L., Arkai, P., Rainer, T.M., Balogh, K., Appel, E. (2007). Temperature and age constraints on the metamorphism of the Tethyan Himalaya in Central Nepal: A multidisciplinary approach. *Journal of Asian Earth Sciences*, 30(1), 113–130.

DeCelles, P.G., Kapp, P., Gehrels, G.E., Ding, L. (2014). Paleocene–Eocene foreland basin evolution in the Himalaya of southern Tibet and Nepal: Implications for the age of initial India–Asia collision. *Tectonics*, 33(5), 824–849.

Dèzes, P.J., Vannay, J.C., Steck, A., Bussy, F., Cosca, M. (1999). Synorogenic extension: Quantitative constraints on the age and displacement of the Zanskar shear zone (northwest Himalaya). *Geological Society of America Bulletin*, 111, 364–374.

Draganits, E. and Noffke, N. (2004). Siliciclastic stromatolites and other microbially induced sedimentary structures in an Early Devonian barrier-island environment (Muth Formation, NW Himalayas). *Journal of Sedimentary Research*, 74(2), 191–202.

Draganits, E., Mawson, R., Talent, J.A., Krystyn, L. (2002). Lithostratigraphy, conodont biostratigraphy and depositional environment of the Middle Devonian (Givetian) to Early Carboniferous (Tournaisian) Lipak Formation in the Pin Valley of Spiti (NW India). *Rivista Italiana di Paleontologia e Stratigrafia*, 108, 7–35.

Dunkl, I., Antolin, B., Wemmer, K., Rantitsch, G., Kienast, M., Montomoli, C., Ding, L., Carosi, R., Appel, E., El Bay, R. et al. (2011). Metamorphic evolution of the Tethyan Himalayan flysch in SE Tibet. In *Growth and Collapse of the Tibetan Plateau*, Gloaguen, R. and Ratschbacher, L. (eds). Geological Society of London Special Publication, London.

Dürr, S.B. and Gibling, M.R. (1994). Early Cretaceous volcaniclastic and quartzose sandstones from north central Nepal: Composition, sedimentology and geotectonic significance. *Geologische Rundschau*, 83(1), 62–75.

Epard, J-L. and Steck, A. (2004). The Eastern prolongation of the Zanskar Sher Zone (Western Himalaya). *Ecolgae Geol. Helv.*, 97, 193–212.

Epard, J.-L. and Steck, A. (2008). Structural development of the Tso Morari ultra-high pressure nappe of the Ladakh Himalaya. *Tectonophysics*, 451, 242–264.

Frank, W. and Fuchs, G.R. (1970). Geological investigations in west Nepal and their significance for the geology of the Himalayas. *Geologische Rundschau*, 59(2), 552–580.

Fuchs, G. (1977). The geology of the Karnali and Dolpo regions, western Nepal. *Jahrbuch der Geologischen Bundesanstalt Wien*, 120, 165–217.

Gaetani, M. and Garzanti, E. (1991). Multicyclic history of the northern India continental margin (NW Himalaya). *American Association of Petroleum Geologists Bulletin*, 75, 1427–1446.

Gaetani, M., Casnedi, R., Fois, E., Garzanti, E., Jadoul, F., Nicora, A., Tintori, A. (1986). Stratigraphy of the Tethys Himalaya in Zanskar, Ladakh. *Rivista Italiana di Paleontologia e Stratigrafia*, 91(4), 443–478.

Gaetani, M., Garzanti, E., Tintori, A. (1990). Permo-Carboniferous stratigraphy in SE Zanskar and NW Lahul (NW Himalaya, India). *Eclogae Geologicae Helvetiae*, 83(1), 143–161.

Gansser, A. (1980). The significance of the Himalayan suture zone. *Tectonophysics*, 62(1–2), 37–52.

Garzanti, E. (1992). Stratigraphy of the Early Cretaceous Giumal Group (Zanskar Range, Northern India). *Rivista italiana di Paleontologia e Stratigrafia*, 97(3–4), 485–509.

Garzanti, E. (1993a). Himalayan ironstones, "superplumes", and the breakup of Gondwana. *Geology*, 21, 105–108.

Garzanti, E. (1993b). Sedimentary evolution and drowning of a passive margin shelf (Giumal Group; Zanskar Tethys Himalaya, India): Palaeoenvironmental changes during final break-up of Gondwanaland. In *Himalayan Tectonics*, Treloar, P.J. and Searle, M.P. (eds). Geological Society of London, Special Publication, London.

Garzanti, E. (1999). Stratigraphy and sedimentary history of the Nepal Tethys Himalayan passive margin. *Journal of Asian Earth Sciences*, 17(5–6), 805–827.

Garzanti, E. and Brignoli, G. (1989). Low temperature metamorphism in the Zanskar sedimentary nappes (NW Himalaya, India). *Eclogae Geologicae Helvetiae*, 82(2), 669–684.

Garzanti, E. and Hu, X. (2015). Latest Cretaceous Himalayan tectonics: Obduction, collision or Deccan-related uplift? *Gondwana Research*, 28(1), 165–178.

Garzanti, E. and Pagni Frette, M. (1991). Stratigraphic succession of the Thakkhola region (Central Nepal) – Comparison with the northwestern Tethys Himalaya. *Rivista Italiana di Paleontologia e Stratigrafia*, 97(1), 3–26.

Garzanti, E. and Sciunnach, D. (1997). Early Carboniferous onset of Gondwanian glaciation and Neo-Tethyan rifting in Southern Tibet. *Earth and Planetary Science Letters*, 148, 359–365.

Garzanti, E., Casnedi, R., Jadoul, F. (1986). Sedimentary evidence of a Cambro-Ordovician orogenic event in the northwestern Himalaya. *Sedimentary Geology*, 48, 237–265.

Garzanti, E., Baud, A., Mascle, G. (1987). Sedimentary record of the northward flight of India and its collision with Eurasia (Ladakh Himalaya, India). *Geodinamica Acta*, (1/4–5), 297–312.

Garzanti, E., Haas, R., Jadoul, F. (1989). Ironstones in the Mesozoic passive margin sequence of the Tethys Himalaya (Zanskar, Northern India): Sedimentology and metamorphism. In *Phanerozoic Ironstones*, Young, T.P. and Taylor, W.E.G. (eds). Geological Society of London, Special Publication, London.

Garzanti, E., Nicora, A., Tintori, A. (1992). Paleozoic to Early Mesozoic stratigraphy and sedimentary evolution of central Dolpo (Nepal Himalaya). *Rivista Italiana di Paleontologia e Stratigrafia*, 98(3), 271–298.

Garzanti, E., Gorza, M., Martellini, L., Nicora, A. (1994a). Transition from diagenesis to metamorphism in the Paleozoic to Mesozoic succession of the Dolpo-Manang Synclinorium and Thakkola Graben (Nepal Tethys Himalaya). *Eclogae Geologicae Helveticae*, 87(2), 613–632.

Garzanti, E., Nicora, A., Tintori, A. (1994b). Triassic stratigraphy and sedimentary evolution of the Annapurna Tethys Himalaya (Manang area, Central Nepal). *Rivista Italiana di Paleontologia e Stratigrafia*, 100(2), 195–226.

Garzanti, E., Nicora, A., Tintori, T., Sciunnach, D., Angiolini, L. (1994c). Late Paleozoic stratigraphy and petrography of the Thini Chu Group (Manang, Central Nepal): Sedimentary record of Gondwana glaciation and rifting of Neotethys. *Rivista Italiana di Paleontologia e Stratigrafia*, 100(2), 155–194.

Garzanti, E., Jadoul, F., Nicora, A., Berra, F. (1995). Triassic of Spiti (Tethys Himalaya; N India). *Rivista Italiana di Paleontologia e Stratigrafia*, 101(3), 267–300.

Garzanti, E., Angiolini, L., Sciunnach, D. (1996a). The mid-Carboniferous to lowermost Permian succession of Spiti (Po Group and Ganmachidam Fm.; Tethys Himalaya, Northern India): Gondwana glaciation and rifting of Neo-Tethys. *Geodinamica Acta*, 9(2), 78–100.

Garzanti, E., Angiolini, L., Sciunnach, D. (1996b). The Permian Kuling Group (Spiti, Lahaul and Zanskar; NW Himalaya): Sedimentary evolution during rift/drift transition and initial opening of Neo-Tethys. *Rivista Italiana di Paleontologia e Stratigrafia*, 102(2), 175–200.

Garzanti, E., Angiolini, L., Brunton, H., Sciunnach, D., Balini, M. (1998a). The Bashkirian "Fenestella shales" and the Moscovian "Chaetetid shales" of the Tethys Himalaya (South Tibet, Nepal and India). *Journal of Asian Earth Sciences*, 16, 119–141.

Garzanti, E., Nicora, A., Rettori, R. (1998b). Permo-Triassic boundary and Lower to Middle Triassic in South Tibet. *Journal of Asian Earth Sciences*, 16, 143–157.

Garzanti, E., Le Fort, P., Sciunnach, D. (1999). First report of Lower Permian basalts in South Tibet: Tholeiitic magmatism during break-up and incipient spreading in Neotethys. *Journal of Asian Earth Sciences*, 17, 533–546.

Gibling, M.R., Gradstein, F.M., Kristiansen, I.L., Nagy, J., Sarti, M., Wiedmann, J. (1994). Early Cretaceous strata of the Nepal Himalayas: Conjugate margins and rift volcanism during Gondwanan breakup. *Journal of the Geological Society of London*, 151, 269–290.

Hayden, H.H. (1904). The geology of Spiti, with parts of Bashar and Rupshu. Geological Survey of India, Memoir 36, 1–129.

Heim, A. and Gansser, A. (1939). Central Himalaya. Geological observations of the Swiss expedition 1936. *Mémoires de la Société Helvétique des Sciences Naturelles*, 73, 1–246.

Henderson, A.L., Najman, Y., Parrish, R., BouDagher-Fadel, M., Barford, D., Garzanti, E., Andò, S. (2010). Geology of the Cenozoic Indus Basin sedimentary rocks: Paleoenvironmental interpretation of sedimentation from the western Himalaya during the early phases of India–Eurasia collision. *Tectonics*, 29(6), TC6015, doi:10.1029/2009TC002651.

Herren, E. (1987). Zanskar shear zone: Northeast-southwest extension within the Higher Himalayas (Ladakh, India). *Geology*, 15, 409–413.

Hu, X., Jansa, L., Wang, C. (2008). Upper Jurassic–Lower Cretaceous stratigraphy in south-eastern Tibet: A comparison with the western Himalayas. *Cretaceous Research*, 29(2), 301–315.

Hu, X., Jansa, L., Chen, L., Griffin, W.L., O'Reilly, S.Y., Wang, J. (2010). Provenance of Lower Cretaceous Wölong Volcaniclastics in the Tibetan Tethyan Himalaya: Implications for the final breakup of Eastern Gondwana. *Sedimentary Geology*, 223, 193–205.

Hu, X., Sinclair, H.D., Wang, J., Jiang, H., Wu, F. (2012). Late Cretaceous-Palaeogene stratigraphic and basin evolution in the Zhepure Mountain of southern Tibet: Implications for the timing of India-Asia initial collision. *Basin Research*, 24(5), 520–543.

Hu, X., Garzanti, E., An, W., Hu, X.F. (2015a). Provenance and drainage system of the Early Cretaceous volcanic detritus in the Himalaya as constrained by detrital zircon geochronology. *Journal of Palaeogeography*, 4(1), 85–98.

Hu, X., Garzanti, E., Moore, T., Raffi, I. (2015b). Direct stratigraphic dating of India-Asia collision onset at the Selandian (middle Paleocene, 59±1 Ma). *Geology*, 43(10), 859–862.

Jadoul, F., Garzanti, E., Fois, E. (1990). Upper Triassic – Lower Jurassic stratigraphy and palaeogeographic evolution of the Zanskar Tethys Himalaya (Zangla Unit). *Rivista Italiana di Paleontologia e Stratigrafia*, 95(4), 351–396.

Jadoul, F., Berra, F., Garzanti, E. (1998). The Tethys Himalayan passive margin from Late Triassic to Early Cretaceous (South Tibet). *Journal of Asian Earth Sciences*, 16, 173–194.

Li, J., Hu, X., Garzanti, E., An, W., Wang, J. (2015). Paleogene carbonate microfacies and sandstone provenance (Gamba area, South Tibet): Stratigraphic response to initial India–Asia continental collision. *Journal of Asian Earth Sciences*, 104, 39–54.

Li, J., Hu, X., Garzanti, E., BouDagher-Fadel, M. (2017). Shallow-water carbonate responses to the Paleocene–Eocene thermal maximum in the Tethyan Himalaya (southern Tibet): Tectonic and climatic implications. *Palaeogeography, Palaeoclimatology, Palaeoecology*, 466, 153–165.

Meng, Z., Wang, J.G., Garzanti, E., Han, Z., Chen, G. (2021). Late Triassic rifting and volcanism on the northeastern Indian margin: A new phase of Neo-Tethyan seafloor spreading and its paleogeographic implications. *Palaeogeography, Palaeoclimatology, Palaeoecology*, 570, 110367.

Montomoli, C., Iaccarino, S., Antolin, B., Appel, E., Carosi, R., Dunkl, I., (2017). Tectono-metamorphic evolution of the Tethyan sedimentary sequence (Himalayas, SE Tibet). *Italian Journal of Geosciences*, 136(1), 73–88.

Myrow, P.M., Hughes, N.C., Searle, M.P., Fanning, C.M., Peng, S.C., Parcha, S.K. (2009). Stratigraphic correlation of Cambrian–Ordovician deposits along the Himalaya: Implications for the age and nature of rocks in the Mount Everest region. *Geological Society of America Bulletin*, 121(3–4), 323–332.

Najman, Y., Appel, E., Boudagher-Fadel, M., Bown, P., Carter, A., Garzanti, E., Godin, L., Han, J., Liebke, U., Oliver, G. et al. (2010). Timing of India–Asia collision: Geological, biostratigraphic, and palaeomagnetic constraints. *Journal of Geophysical Research*, 115, B12416. doi:10.1029/2010JB007673.

Najman, Y., Jenks, D., Godin, L., Boudagher-Fadel, M., Millar, I., Garzanti, E., Horstwood, M., Bracciali, L. (2017). The Tethyan Himalayan detrital record shows that India–Asia terminal collision occurred by 54 Ma in the Western Himalaya. *Earth and Planetary Science Letters*, 459, 301–310.

Nicora, A. and Garzanti, E. (1997). The Permian/Triassic boundary in the Central Himalaya. *Albertiana*, 19, 47–51.

Nicora, A., Garzanti, E., Fois, E. (1987). Evolution of the Tethys Himalaya continental shelf during Maastrichtian to Paleocene (Zanskar, India). *Rivista Italiana di Paleontologia e Stratigrafia*, 92(4), 439–496.

Patriat, P., Segoufin, J., Schlich, R., Goslin, J., Auzende, J.-M., Beuzart, P., Bonnin, J., Olivet, J.-L. (1982). Les mouvements relatifs de l'Inde, de l'Afrique et de l'Eurasie. The relative movements of India, Africa and Eurasia. *Bulletin de la Société Géologique de France*, S7-24(2), 363–373.

Premoli Silva, I., Garzanti, E., Gaetani, M. (1991). Stratigraphy of the Chikkim and Fatu La Formations in the Zangla and Zumlung Units (Zanskar Range, India) with comparisons to the Thakkhola region (central Nepal): Mid-Cretaceous evolution of the Indian passive margin. *Rivista Italiana di Paleontologia e Stratigrafia*, 97, 511–564.

von Rad, U., Exon, N.F., Haq, B.U. (1992). Rift-to-drift history of the Wombat Plateau, Northwest Australia: Triassic to Tertiary Leg 122 Results. In *Proceedings of the Ocean Drilling Program, Scientific Results*, von Rad, U., Haq, B.U. et al. (eds), Vol. 122(46), 765–800.

Sakagami, S., Sciunnach, D., Garzanti, E. (2006). Late Paleozoic and Triassic bryozoans from the Tethys Himalaya (N India, Nepal and S Tibet). *Facies*, 52(2), 279–298.

Schill, E., Appel, E., Gautam, P. (2002). Towards pyrrhotite/magnetite geothermometry in low-grade metamorphic carbonates of the Tethyan Himalayas (Shiar Khola, central Nepal). *Journal of Asian Earth Science*, 20, 195–201.

Schneider, C. and Masch, L. (1993). The metamorphism of the Tibetan series from the Manang Area, Marsyandi Valley, Central Nepal, 74. In *Himalayan Tectonics*, Treloar, P.J. and Searle, M.P. (eds). Geol. Soc., Special Publication, London.

Sciunnach, D. and Garzanti, E. (1996). Sedimentary record of Late Paleozoic rift and break-up in Northern Gondwana (Thini Chu Group and Tamba-Kurkur Fm.; Dolpo Tethys Himalaya, Nepal). *Geodinamica Acta*, 9(1), 41–56.

Sciunnach, D. and Garzanti, E. (1997). Detrital chromian spinels record tectono-magmatic evolution from Carboniferous rifting to Permian spreading in Neotethys (India, Nepal and Tibet). In *From Rifting to Drifting in Present-day and Fossil Ocean Basins*, Messiga, B. and Tribuzio, R. (eds). Ofioliti, 22(1), 101–110.

Sciunnach, D. and Garzanti, E. (2012). Subsidence history of the Tethys Himalaya. *Earth-Science Reviews*, 111(1–2), 179–198.

Searle, M.P. (1999). Extensional and compressional faults in the Everest-Lhotse Massif, Khumbu Himalaya, Nepal. *Journal of the Geological Society*, 156, 227–240.

de Sigoyer, J., Chavagnac, V., Blichert-Toft, J., Villa, I.M., Luais, P., Guillot, S., Cosca, M., Mascle, G. (2000). Dating the Indian continental subduction and collisional thickening in the northwest Himalaya: Multichronology of the Tso Morari eclogites. *Geology*, 28, 487–490.

Singh, A.K., Chung, S.L., Bikramaditya, R., Lee, H.Y., Khogenkumar, S. (2020). Zircon U–Pb geochronology, Hf isotopic compositions, and petrogenetic study of Abor volcanic rocks of Eastern Himalayan Syntaxis, Northeast India: Implications for eruption during breakup of Eastern Gondwana. *Geological Journal*, 55(2), 1227–1244.

Spring, L. and Crespo-Blanc, A. (1992). Nappe tectonics, extension, and metamorphic evolution in the Indian Tethys Himalaya (Higher Himalaya, SE Zanskar and NW Lahul). *Tectonics*, 11(5), 978–989.

Steck, A. (2003). Geology of the NW Indian Himalaya. *Eclogae Geologicae Helvetiae*, 96, 147–196.

Steck, A., Epard, J.-L., Vannay. J.-C., Hunziker, J., Girard, M., Morard, A., Robyr, M. (1998). Geological transect across the Tso Morari and Spiti areas: The nappe structures of the Tethys Himalaya. *Eclogae Geologicae Helvetiae*, 91, 103–121.

Stoliczka, A. (1866). Summary of the geological observations during a visit to the provinces Rupshu Karnag, South Ladakh, Zanskar, Sumdo and Dras of western Tibet. Geological Survey of India, Memoir 5, 337–354.

Vannay, J.-C. and Spring, L. (1993). Geochemistry of the continental basalts within the Tethyan Himalaya of Lahul-Spiti and SE Zanskar northwest India. In *Himalayan Tectonics*, Treloar, P.J. and Searle, M.P. (eds). Geological Society of London, Special Publication, London.

Wang, J.G., Wu, F.Y., Garzanti, E., Hu, X., Ji, W.Q., Liu, Z.C., Liu, X.C. (2016). Upper Triassic turbidites of the northern Tethyan Himalaya (Langjiexue Group): The terminal of a sediment-routing system sourced in the Gondwanide Orogen. *Gondwana Research*, 34, 84–98.

Willems, H., Zhou, Z., Zhang, B.G., Gräfe, K.U. (1996). Stratigraphy of the Upper Cretaceous and lower Tertiary strata in the Tethyan Himalayas of Tibet (Tingri area, China). *Geologische Rundschau*, 85(4), 723–754.

Zhu, D.C., Chung, S.L., Mo, X.X., Zhao, Z.D., Niu, Y., Song, B., Yang, Y.H. (2009). The 132 Ma Comei-Bunbury large igneous province: Remnants identified in present-day southeastern Tibet and southwestern Australia. *Geology*, 37(7), 583–586.

PART 2

Greater Himalayan Crystalline Complex

4

High-Pressure and Ultra-High-Pressure Units in the Himalaya

Julia DE SIGOYER and Stéphane GUILLOT

University of Grenoble Alpes, France

4.1. Introduction

The discovery of high-pressure and ultra-high-pressure (HP–UHP) metamorphic rocks in the Himalaya overturns the concepts of the formation of this mountain range, which until the 1980s was considered as the archetype of the collisional belt. Indeed, HP–UHP metamorphic rocks preserved in mountain ranges are important petrological markers of major geodynamic processes such as oceanic subduction (Ernst 1973), continental subduction (Chopin 1984) and continent–continent collision.

High- to ultra-high-pressure (HP–UHP) metamorphic rocks occur in different units of the Himalayan range (from the Suture Zone in the north to the Higher Himalayan Crystallines in the south). They are of various ages, natures and metamorphic conditions and therefore provide access to different key stages of the formation of the Himalayan range from Neotethys oceanic subduction to the India–Asia continental collision.

Himalaya, Dynamics of a Giant 2,
coordinated by Rodolphe CATTIN and Jean-Luc EPARD.
© ISTE Ltd 2023.

Three types of high-pressure metamorphic rocks are observed in the Himalaya: (i) the oldest ones are the blueschists facies and low-temperature (LT) eclogitic facies rocks in the suture zone related to Tethyan oceanic subduction, (ii) the LT eclogitic facies rocks of high- to ultra-high pressure present in the Indian continental margin just south of the Indus Tsangpo Suture Zone (ITSZ). Some of them are middle temperature eclogite, and the younger granulitized eclogites of high pressure and high temperature (HP–HT) observed in the continental units of the Higher Himalayan Crystallines (HHC) far from the suture zone that are related to the important crustal thickening during collision. In this chapter, we propose a review of these different HP–UHP metamorphic rocks, pointing out the constraints they bring on the formation of the Himalayan range.

4.2. High pressure rocks in the suture zone (witnesses of the oceanic subduction)

The blueschists and lawsonite (HP very LT metamorphic mineral) bearing eclogite in the Indus Tsang Po suture zone (ITSZ) are quite rare in the Himalaya contrary to other Cenozoic belts. They are mostly observed in the north-western Himalaya (Shangla) Pakistan (Shams 1972; Franck et al. 1977) and Ladakh (NW India) Sapi–Shergol: (Honegger et al. 1989) (Figure 4.1) or in the eastern Himalaya in the Indo-Burmese Ranges (Nagaland Ophiolite Complex): (Chatterjee and Ghose 2010), Chin Hill Ophiolite: where they are interpreted as the eastern extension of the ITSZ.

In both areas (NW Himalaya and E Himalaya), the blueschists and LT eclogites are adjacent to the ophiolitic unit of the ITSZ (suture zone). They represent the paleo-accretionary prism associated with the northward subduction of the Neotethyan ocean beneath the Asian continental margin (e.g. Mahéo et al. 2006; Guillot et al. 2008). The rare high-pressure/low-temperature (HP–LT) rocks represent the deepest part of the accretionary wedge in front of the subduction trench. These rocks are then crucial to constrain the evolution of the closure of the Neotethyan ocean (Guillot et al. 2008; Groppo et al. 2016; O'Brien 2019) (see Volume 2 – Chapter 2).

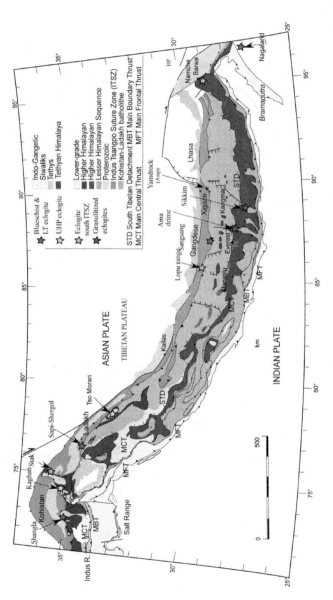

Figure 4.1. *Simplified geological map of the Himalayan orogen showing the location of (i) the blueschist facies and LT eclogitic rocks in the Indus–Tsangpo suture zone (ITSZ) blue stars; (ii) the LT HP to UHP unit of the Indian continental margin light yellow stars, and the MT eclogite south of the ITSZ; (iii) the HT eclogite observed north of Higher Himalaya red stars (modified from Guillot et al. (2008). For a color version of this figure, see www.iste.co.uk/cattin/himalaya2.zip)*

4.2.1. *The Shapi–Shergol blueschists (Ladakh)*

The Shapi–Shergol blueschists in the suture zone of the NW Himalaya (Ladakh) (Figure 4.1) present volcanoclastic protolith, rich in mafic metamorphosed rocks that are interpreted as an ophiolitic melange typical of a paleo-accretionary prism (Honegger et al. 1989). The metamorphic paragenesis of these blueschists can vary from one bloc of rocks to another depending on their protolith. Most of metabasic blocs contain fresh lawsonite, glaucophane, phengite and minor clinopyroxene (omphacite). The associated metasediments contain carpholite, lawsonite or zoisite, glaucophane, phengite and garnet. All these minerals are stable under low-temperature and high-pressure conditions. Pressure peak conditions are estimates ca. 470°C, 1.9 GPa (Groppo et al. 2016). The preservation of fresh lawsonite in these rocks without retrogression into zoisite or epidote during exhumation implies that the exhumation occurred along a very low thermal gradient (<7–9°C/Km) only possible in a subduction context. K–Ar and Ar/Ar datings of whole-rocks and glaucophane suggest an age of ca. 100–130 Ma for the high-pressure metamorphism (Honegger et al. 1989; Mahéo et al. 2004).

4.2.2. *The Shangla Blueschists (Pakistan)*

The Shangla Blueschists (Pakistan) present similar lithological characteristic than the Shapi–Shergol ones. They only differ by the absence of lawsonite in the paragenesis. The peak PT conditions are estimated at 350–520°C and 0.9–1.2 GPa by Iqbal et al. (2020). The prograde P–T path is almost parallel to the retrograde path, implying an exhumation into the paleo-accretionary prism during the subduction of the Neotethys. The peak of metamorphism is dated using K–Ar, Ar/Ar and Rb–Sr radiometric techniques for Na amphiboles and phengites at ca. 80 Ma (Anczkiewicz et al. 2000).

4.2.3. *The Sangsang Blueschist*

The Sangsang Blueschists are located in the ITSZ of the South Central Tibet at the longitude of the Everest. Ophiolitic melange including metabasic rocks and metasediments were metamorphosed under blueschist facies conditions <540°C. Wang et al. (2017) proposed a protolith age at 149 Ma (U/Pb on zircon) and a metamorphic age at 60–63 Ma (Ar/Ar on amphibole and phengite).

4.2.4. The Indo-Burmese Blueschists

In the Indo-Burmese ranges, the southeastern lateral continuity of the ITSZ, varieties of HP–LT metabasic rocks are described, including lawsonite bearing blueschists, epidote blueschists and lawsonite bearing eclogites. The different portions of the suture zone present variable peak P–T conditions ranging from ∼340 °C, ∼11.5 kbar for the lawsonite blueschists, 540 ± 35°C, 14.4 ± 2 kbar for the epidote blueschists to 580–610°C and 17–20 kbar for the lawsonite bearing eclogites (Chatterjee and Ghose 2010).

Therefore, from west to east, the occurrence of lawsonite bearing blueschist and eclogites reveals the presence of the accretionary wedge of a cold subduction of the Neotethys ocean northward under the Asian margin. This cold subduction is dated between 130 and 100 Ma for the first group and 65 Ma for the second group, that is, before the Indian–Asian collision.

4.3. Continental high-pressure (HP) to ultra-high-pressure (UHP) metamorphism of the Indian margin (continental subduction) located next to the Indus Tsangpo Suture Zone

The UHP metamorphic rocks are defined by the occurrence of coesite or diamond (Chopin 1984). Coesite is the polymorph of quartz at a high pressure above 2.5 GPa and ca. 600–700°C. Coesite is not stable at a crustal depth, even under the Tibet at 80 km as the temperature is too high. Thus, neither the oceanic nor the continental crust contains coesite. The occurrence of coesite in a rock involves that it has crystallized at a mantle depth (>80 km) (Figure 4.2). Therefore, the discovery of UHP mineral such as coesite in upper crust continental rocks implies that the continental lithosphere including the crust went down very deep to the mantle depth.

Only two occurrences of UHP metamorphism affecting the Indian continental margin were discovered in the NW Himalaya: they are observed in the Kaghan Valley (Pakistan) and in the Tso Morari dome (Ladakh) (Figure 4.1) (Pognante and Spencer 1991; Guillot et al. 1997, 2008; de Sigoyer et al. 2000; O'Brien et al. 2001; Sachan et al. 2004). They are separated by 500 km, both are located south of the ITSZ. They share many characteristics. They both show anticlinal crystalline domes that represent the basement of the Indian continental margin with its Permo-Triassic sedimentary cover. Both the basement and the cover were metamorphosed under UHP metamorphic conditions. The anticlinal domes are surrounded by normal shear zones.

4.3.1. The Kaghan unit

In the Kagan Valley, the UHP–HP unit lies just south of the Kohistan arc with local slivers of ITSZ in between. It spread on a surface of less than 1,000 km^2. The Proterozoic basement consists of metapelite, metapsamite and orthogneisses. It is overlaid by a Paleozoic cover first with clastic and then carbonate-rich sediments containing dykes and lenses of metabasic rocks and felsic intrusions. The basics dyke and sill are Permian Panjal trap with a subalkaline to alkaline tholeiitic composition (Pognante and Spencer 1991). Many U/Pb analyses on zircon cores date the protolith crystallization during the Permian (Kaneko et al. 2003). In these metabasic rocks, the eclogite paragenesis consists of coesite as inclusion in omphacite or garnet or zircon + phengite + rutile ± glaucophane ± zoisite magnesite/dolomite (O'Brien et al. 2001; Wilke et al. 2010). Na–Ca amphibole associated with Na–clinopyroxene+albite appears in the symplectitic reaction of the omphacites. Titanite appears at the expense of rutiles during the retrogression under amphibolitic conditions. The later metamorphic paragenesis contains chlorite with albite that testify to pervasive greenschist overprint. Most of the metasediments contain garnet, kyanite and staurolite and are also retrogressed under greenschist facies conditions.

Peak metamorphic conditions are estimated at 2.5–3.2 GPa and 700 ±70°C (Lombardo and Rolfo 2000; Wilke et al. 2010). The post-peak growth of glaucophane requires cooling to 580–630°C (at 1.0–1.7 GPa) during the initial exhumation (Wilke et al. 2010) followed by a stage of heating to 650–720°C (at 1.0–1.2 GPa), and then a decompression towards greenschist conditions.

In the Kaghan Valley, the age of the UHP peak is well constrained at 46.2 ± 0.7 Ma (2σ) based on U–Pb dating of zircon rims that contain coesite inclusions (Kaneko et al. 2003). This age is confirmed by other radiometric methods applied on gneiss or on eclogites (see O'Brien (2019), for a complete review). The exhumation is dated at 39–42 Ma by Ar/Ar dating on amphibole. Apatite fission tracks finally give 24.5 ± 3.7 Ma for the end of the exhumation.

High-Pressure and Ultra-High-Pressure Units in the Himalaya 89

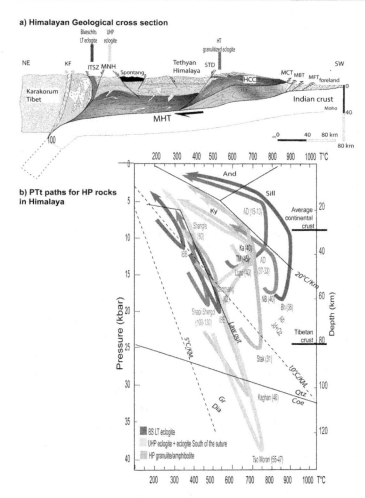

Figure 4.2. a) Theoretical geological NS cross-section across the Himalayan belt showing the location of the different HP rocks. NHM (North Himalayan Massif), STD South Tibetan Detachment, MCT Main Central Thrust, MBT Main Boundary thrust, MFT Main Frontal Thrust, (modified from Guillot et al. (2008)). b) Synthesis of the PTt paths for all the HP rocks described in the Himalaya. The blueschists and LT eclogite in the ITSZ are in blue. The UHP eclogites in the North Himalayan massif, south of the ITSZ are in light yellow, MT eclogites located south of the ITSZ are in dark yellow and brown; granulitized eclogites of the HHC are in red. IBB, IBE Indo-Burmese Blueschists, eclogites, NB Namche Barwa, AM Ama Drime, Bh Bhutan. See the text for the reference of the PTt paths. For a color version of this figure, see www.iste.co.uk/cattin/himalaya2.zip

4.3.2. The Tso Morari UHP unit

The Tso Morari unit extends south of the Indus Suture Zone along 5,000 km^2 (Figure 4.1). The Tso Morari unit presents a domal shape. It is separated from the ITSZ by a normal shear zone where HP serpentinites originating from the mantle wedge were discovered (Guillot et al. 2001). Its southern boundary presents also a normal shear in between the Tso Morari and the Zanskar unit (Tethyan margin). The Tso Morari dome forms part of North Himalayan dome that lies just south of the suture. It has been considered as a nappe system by Steck et al. (1993), and forms part of the North Himalayan massif.

The metagranitic and orthogneissic basement in the Tso Morari is Ordovician in age, and dated at 479 ± 2 Ma by U–Pb on zircon (de Sigoyer 1998). A probable Neoproterozoic basement is also observed as it is overlaid by a Late Precambrian to lower Paleozoic sedimentary cover of paragneiss and micaschists (de Sigoyer et al. 2004). Eclogitic facies is observed in the metabasic lenses or dykes within the basement and in the metasedimentary cover (Figure 4.3). It is more difficult to identify it in the orthogneiss. The protolith of Kaghan and Tso Morari mafic eclogites is dated at about 280 Ma (Rajkumar 2015; Rehman et al. 2016), and is attributed to the eruption of Panjal Traps as observed elsewhere on the Indian continental margin. The lithological association observed in the Tso Morari unit is very typical of the distal Indian continental margin block (de Sigoyer et al. 2004).

Berthelsen (1953) was the first to describe the metamorphism of the eclogitic basic lenses (even if he did not name them as eclogites). All the lithologies (basement metasediments and metabasic rocks) underwent UHP–LT conditions (Figure 4.3). The metabasic lenses were the subject of numerous studies. Eclogitic paragenesis consists of garnet core inclusions rich, omphacite, phengite, rutile, quartz, zoisite, Na amphibole (such as glaucophane), dolomite is observed in the garnet core, calcite in the garnet rim and trace of zircon (Figure 4.3). Inclusions of coesite were suspected by de Sigoyer et al. (1997) and described in the garnet rim and omphacite by Sachan et al. (2004) (Figure 4.3) suggesting that garnet rim grew during the peak of metamorphism. Wilke et al. (2015) propose that dolomite grew at the expense of magnesite and aragonite.

The cores of garnet associated with their numerous inclusions of calcic and sodic-calcic amphiboles, chloritoid, epidote, paragonite, calcite, quartz and zircon allow us to estimate prograde P–T conditions at 2.15 ± 0.15 GPa and 535 ± 15°C (Figure 4.2).

The mantle of garnet is almost inclusion free but can present coesite omphacite, paragonite, baroisite, epidote, phengite, talc, epidote/allanite, rutile, calcite, zircon and quartz. They represent the peak of pressure for these eclogites estimated at 2.55–2.75 GPa and 630–645°C (Saint-Onge et al. 2013). Mukheerjee et al. (2003) proposed very high P–T conditions >3.7 GPa 750°C. Considering the reaction dolomite = magnesite + aragonite, Wilke et al. (2015) propose much higher P–T conditions at 4.4–4.8 GPa at 560–760°C in the field of diamond. Nevertheless, neither diamond nor aragonite was clearly proved in the Tso Morari unit.

The first part of the retrogression observed in the matrix is characterized by the occurrence of glaucophane (stable below 1.7 GPa and 550°C) in addition to the previous paragenesis (unless the coesite). The beginning of exhumation occurred then in the eclogitic field at low temperature (de Sigoyer et al. 1997). Then, Ca amphibole developed around the garnet, sodic augite–plagioclase symplectites in the matrix, pyrite, titanite after rutile. This amphibolitic paragenesis reflects a second part of exhumation associated with significant heating at ∼0.9–1.2 GPa, 650–720°C (de Sigoyer et al. 1997; Saint-Onge et al. 2013). Such an increase of temperature during the return path is also observed in the metasediments with the occurrence of staurolite and kyanite (Guillot et al. 1997; Mukheerjee and Sachan 2001). The end of the exhumation occurred under greenschist facies conditions up to the surface (de Sigoyer et al. 1997; Saint-Onge et al. 2013) (Figure 4.2).

The age of eclogitic phase including prograde path, peak of UHP metamorphism and beginning of cold decompression in eclogites and host rocks is dated in between 47 and 55 Ma using different methods (de Sigoyer et al. 2000; Leech et al. 2005; Saint-Onge et al. 2013). The exhumation associated with heating up to 10 km depth (shallow crustal levels) occurred in between 45 and 30 Ma based on U–Pb on retrogressed zircon, Ar–Ar on phengites, apatite fission track and zircon U–He–Th ages as well as from diffusion modeling of zoning in garnet (de Sigoyer et al. 2000; Wilke et al. 2015).

Figure 4.3. *(a) Photo of a metamafic lens of eclogite into the Puga gneiss and metasediments of the Tso Morari massif. (b) Photomicrograph plane polar of a metapelite showing the eclogitic paragenesis with poikilitic garnet, containing the inclusion of jadeite, the matrix is composed of quartz phengite glaucophane and chloritoide. (c) Photomicrograph under crossed nikols of the eclogitic foliation in the core of a metabasic lens, underlined by the dynamic recrystallization of omphacite (omph) and garnet (grt). (d) Photomicrograph under the uncrossed polar of Coesite (Coe) inclusion into garnet surrounded by quartz (Qz) at the boundary between the inclusion-rich and inclusion-poor zones of garnet from O'Brien (2019). For a color version of this figure, see www.iste.co.uk/cattin/himalaya2.zip*

4.3.3. Other HP metamorphosed unit south of suture zone in the Indian continental margin

Other high-pressure rocks that belong to the Indian passive margin were identified along the suture zone in Pakistan Stak unit and in southern Tibet in the Lopu Range, 600 km west of Lhasa, and the Namche Barwa in Tibet (Figure 4.1). They present some similarities with the UHP assemblage units but up to now, no UHP was described in these massifs. It should be noted that they underwent higher temperatures relative to Kaghan and Tso Morari units during their exhumation.

Eclogites and high-pressure granulites were described at Stak (Pakistan) just Northeast of the Kaghan Valley, close to the Nanga Parbat Haramosh massif (in the western syntaxis of the Himalaya) along the Main Mantle Thrust (MMT) (Le Fort et al. 1997). This massif consists of Indian continental gneisses, schists, metabasites, with minor marbles. Small lenses of mafic rocks lie in felsic gneisses and form part of the Paleozoic and Mesozoic cover to basement gneisses, as it is observed in the upper HHC further west. The eclogite present similarities with the Kaghan ones with PT conditions at 2.5 GPa and 750°C (Lanari et al. 2013), but the apparition of brown hornblende during the exhumation reflected heating at 0.9–1.6 GPa at 650–700°C (Figure 4.2). Zircon are to small to date precisely the peak of metamorphism without ambiguity and lies in between 158 and 32 Ma. Chen et al. (2022) proposed an age ∼31 Ma, which is at least ∼15 Ma later than the Himalayan UHP metamorphism.

The Lupo range in southern Tibet presents a continental Phanerozoic protolith with Tethyan sedimentological cover. It is separated from the suture by a normal shear zone. High-pressure metamorphism is observed in the metasediments with a paragenesis at Phengite + Garnet + Staurolite + Chlorite.

The pressure and temperature condition determined by Laskowki et al. (2016) are >1.4 GPa at T < 600°C. The garnet crystallization is dated by Lu-Hf method at about 40.4 ± 1.4 Ma, which correspond to the prograde metamorphism. The exhumation up to 25 km depth is dated in between 39 and 34 Ma on phengite with the Ar/Ar single grain method. This unit presents similar structural position and protolith than the Kaghan rocks or the Tso Morari ones located 700 km westward. Up to now, no ultra-high pressure was observed in the Lupo massif and the high-pressure event seems to occur 6 Myr later (Figure 4.2).

The Namche Barwa is located in the eastern virgation of the Himalayan belt (Figure 4.1). It is presented as an anticlinal dome of HHC presenting some structural similarities with the Nanga Parbat the western virgation (Burg et al. 1998). The basement and metasedimentary cover are attributed to the Indian margin. Anatectic gneisses and mafic granulite are observed in the core of the Namche Barwa. In these samples, some relics of garnet, kyanite rutile and quartz attested for a relictual stage at higher pressure from 1.4 to 1.8 GPa and 800 to 890°C (Liu and Zhong 1997; Ding et al. 2001; Zhang et al. 2015). The migmatitic stage, characterized by low-pressure granulite-facies paragenesis is estimated at 0.4–0.6 GPa and 750–850°C (Zhang et al. 2015). The granulitic overprint is dated with zircon, monazite and titanite from 10 Ma to 3 Ma (Ding et al. 2001). The age of high-pressure event could be of about 40 Ma (Ding et al. 2001; Zhang et al. 2015). It was followed by a very rapid exhumation >10 mm/yr with a very rapid cooling >100°/km in the last 5 myr, as deduced by Ar/Ar micas dating and (U–Th)–He on zircon dating and fission tracks on zircon and apatite (Burg et al. 1998).

These three HP north Himalayan massifs, relatively small (<5,000 km^2), represent part of the Indian continental margin that were underthrusted during Late Eocene beneath South Tibet and exhumed along the Main Himalayan Thrust during the Oligocene collision, less than 10 Ma after their burial (Figure 4.2). In contrast, the mantle depth reached by the UHP NW Himalaya massifs imposes a steep subduction of the Indian continental margin, following the Tethyan ocean one during Mid-Eocene time. The Indian margin enters into the subduction zone earlier in the Western Syntaxis along a steeper subduction zone compared to Central and Eastern Himalaya (Zhang et al. 2015).

4.4. Oligocene–Miocene high-pressure, high-temperature metamorphism eclogite with granulite overprint far from the suture zone

HP eclogites strongly overprinted into granulitic facies were described for the first time by Lombardo et al. (1998) in the northern part of the Higher Himalayan Crystallines away from the suture zone (Figures 4.1 and 4.2) in central Nepal (Ama Drime/Arun). Similar HT eclogites were observed in the Eastern Himalaya (Kanchenjunga, North Sikkim and NW Bhutan).

In the Ama Drime/Arun Valley, the eclogite-facies metamorphism is observed in lenses of metabasic into orthogneiss (Kfs + Pl + Qz + Bt) and sillimanite bearing paragneiss with Grt + Sil/Bt or Bt + Sill+ Crd + Ged (Lombardo and Rolfo 2000; Groppo et al. 2007; Wang et al. 2017). The protolith of metabasic lenses is dated with U/Pb on zircon at 1,017.5 ± 9.6 Ma (Wang et al. 2017). The orthogneissic basement is dated at ca. 1,800 Ma zircon (Cottle et al. 2009). The basic rocks are related to Proterozoic rift origin. Such protoliths represent the Indian continental craton, which is different from the distal continental margin signature recognized in the UHP Kaghan or Tso Morari massif.

The eclogitic stage (M1) is attested in the metabasic lenses by the occurrence of omphacite as inclusion in the garnet or zircon. In the matrix, only symplectites of Jadeite poor content of clinopyroxene lamaellas and oligoclase are observed, attesting for the past occurrence of omphacite. The symplectites are locally associated with amphiboles or orthopyroxenes attesting for a strong M2 HP granulitic overprint. The occurrence of corona of orthopyroxene + plagioclase around the garnet suggests an M3 LP granulitic stage. The orthopyroxenes are surrounded by late amphiboles showing a final M4 overprinted under amphibolitic facies conditions. Former phengite has been replaced by biotite in plagioclase (Lombardo and Rolfo 2000; Groppo et al. 2007; Wang et al. 2017) and ilmenite (itself replaced by titanite) has replaced most of the rutile in the matrix.

The initial eclogitic pressure peak condition is estimated at 580°C and 15–16 kbar (Groppo et al. 2007), the M2 HP granulitic overprinted is estimated at >750°C and 8–12 kbar, the M3 LP granulite at 750°C and 4–6 kbar and the last M4 amphibolite-facies overprinted is estimated at (675°C). Higher pressures of 20 kbar at 710 ± 50°C were recently proposed for the eclogitic stage (Wang et al. 2017). Ages of such retrogressed rocks are not easy to retrieve and are strongly discussed. Corrie et al. (2016) argue that the HP pressure predates ca. 21 Ma (Lu–Hf on garnets, while the granulitic overprint in the metabasic lenses is dated at 13–17 Ma (e.g. U–Pb on zircons by SHRIMP; (Groppo et al. 2007; Cottle et al. 2009; Wang et al. 2017)). Kellett et al. (2014) gives garnet Lu–Hf ages of 37.5–33.9 Ma, rather suggesting that the Ama Drime metabasite samples were eclogitized during Eocene, with a Miocene age for the granulite-facies overprint (U–Pb age on zircon 13–15 Ma). The end of exhumation is related to the normal movement along the South Tibetan Detachment fault (Kellett et al. 2014).

Equivalent mafic rocks were observed in Bhutan and Kharta (Eastern Himalaya) with lower P granulitic paragenesis (Crd-, Sil- and Opx-bearing) in paragneisses (Groppo et al. 2007; Chakungal et al. 2010) (Figures 4.1 and 4.2).

4.5. Conclusion

The dynamic of the Himalaya is fast enough to be observed in the same range of different types of HP–UHP rocks: the blueschist and lawsonite bearing eclogites in the suture zone, the UHP unit of the Indian continental margin exhumed just south of the suture zone and the HT eclogites far from the suture zone (Figure 4.2). All these HP rocks are a window on the earliest stage of the Himalayan building, allowing us to document and date the stages related to the oceanic subduction, the continental subduction and the collision.

The blueschists and eclogites in the Indus Tsangpo Suture zone were formed in the accretionary wedge during the Cretaceous subduction of the Neotethys ocean below the Asian active margin. The accretionary wedge was emplaced and stacked in the upper part of the oceanic crust above the subduction zone. Such a type of structure is generally preserved in the young mountain belt. The P–T path of these rocks with a hairpin shape attests for cold conditions during the subduction and during the exhumation of rocks allowing the lawsonite to remain stable all along the exhumation path. Such cold conditions and a hairpin P–T path implies that the exhumation is due and associated with subduction processes (Figures 4.2 and 4.4).

The discovery of a continental unit that contains coesite and other UHP evidence has major implication for the geodynamic of the Himalayan mountain building. At the difference with the accretionary wedge where only small blocks of blueschists or eclogites were metamorphosed at high pressure, the UHP and HP continental units are coherent and the whole unit underwent UHP or HP conditions. The UHP conditions cannot reflect every local overpressure zone as it has been objected in several papers. Most of the time, the eclogite lenses are embedded in metasediments that present a soft rheology. Furthermore, HP minerals are often present as an inclusion on garnet or omphacite. In some eclogite, no deformation is recorded during HP metamorphism magmatic texture that is still preserved, and precluding the local overpressure interpretation of HP to UHP conditions. Therefore, the discovery of HP to UHP north Himalayan massifs south of the suture zone, which represent the leading edge of the Indian continental margin, highlights a stage of continental subduction in between the stage of oceanic subduction and collision in the Himalaya. This northward subduction of the leading edge of the Indian continental margin down to the Asian margin began during the lower Eocene, 55 Ma ago. The presence of coesite in this UHP unit involves that Indian margin going down to mantle depth (>100 km depth) (Figure 4.4).

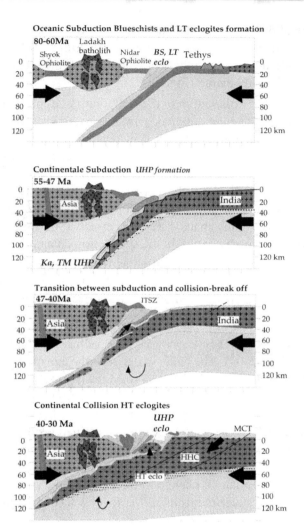

Figure 4.4. *Schematic illustration modified after de Sigoyer et al. (2004) showing the evolution processes of the western Himalaya. (a) The closure of the Tethyan ocean with the formation of blueschists and low temperature eclogite in the ITSZ at 80–60 Ma. (b) Steep subduction and UHP metamorphism of the Indian continental slab at ca. 55–46 Ma. (c) Breakoff of the Neotethyan lithosphere induced fast exhumation of UHP metamorphic rocks along the subduction zone followed by a tilting Indian continental slab at ca. 46–40 Ma. (d) Collision process (ca. 40–25 Ma) with low angle of underthrusting of the Indian continental slab induces the HT HP metamorphism that will become the HHC. For a color version of this figure, see www.iste.co.uk/cattin/himalaya2.zip*

The exhumation of large units such as the Tso Morari and Kaghan units is done in two stages: the first is very rapid (less than 10 Ma) along a cold retrogressed path in the stability field of glaucophane, the second is slower and associated with crustal heating (Figure 4.2):

– The first stage of exhumation of the UHP units takes place in a subduction context. This part of the exhumation occurred very quickly after the subduction of the unit. The subduction process itself involved a return flow path within the serpentinized channel located above the Benioff plane (Guillot et al. 2001). The return velocity is as fast as the subduction up to 11 cm/yr; therefore, the UHP unit can return back into the upper crust less than 10 Ma after being buried. The position of the UHP terranes close to the ITSZ, as the discovery of serpentinite of HP in the shear zone between the Tso Morari and the suture zone also supports an exhumation in the subduction channel.

– The second stage of the exhumation across the crust (from 30 km to the surface) is much more slower and associated with heating, coupling crustal tectonic processes and erosion removal at the surface (de Sigoyer et al. 2004) (Figure 4.2, Figure 4.4).

These two stages of exhumation are typical of most of the UHP unit in the world (Guillot et al. 2009). The delay in between the age of peak metamorphic conditions in NW Himalaya (Kaghan, Tso Morari) and the Central Himalaya (Lupo) may reflect a lateral variation in the subduction dip from west to east during Eocene time. UHP eclogites in the western Himalaya and internal zone were subducted earlier (55–47 Ma) and exhumed faster (45–40 Ma) than the HP rocks in the Central and Eastern Himalaya in the Higher Himalaya Crystallines (buried at 38–15 Ma and exhumed at 25–13 Ma) (Zhang et al. 2015). It may reflect a lateral change in the angle of the subduction dip from West to East and through time from 55 Ma to 35 Ma (Figure 4.4) (Chen et al. 2022).

The discovery of HP HT granulitized eclogite far from the initial subduction zone is also of a major importance. It characterizes the beginning of the strong thickening of the continental crust that is observed under the Himalaya and imaged by passive seismology (Nábělek et al. 2009). Presently, gravity anomaly below the internal part of the Himalaya is interpreted as the present eclogitization of the lower crust due to the overthinking of the crust up to eclogitized conditions (Hetényi et al. 2007). Specific deep seismicity without aftershocks sequence is also interpreted as reflecting dehydration reaction during HT eclogitization of the lower crust in the Himalayan thickened wedge

(Petley-Ragan et al. 2019). Therefore, the characterization of the first evidence of such granulitized eclogites is of major importance. It is important to specify the age for the eclogitic stage Eocene or Miocene and really date the beginning of the collision processes, namely of the crustal stacks. The Higher Himalayan Crystalline is no longer considered a single coherent slab but rather the stacking of different slices of the Indian crust with different histories, separated by ductile shear zones. Slices metamorphosed under HP during the Eocene could be exhumed during the Miocene time; monazite ages in the metapelites indicate a long period of residence at high temperatures (Wang et al. 2017) (Figure 4.4).

4.6. References

Anczkiewicz, R., Burg, J.P., Villa, I.M., Meier, M. (2000). Late Cretaceous blueschists metamorphism in the Indus suture zone, Shangla region, Pakistan Himalaya. *Tectonophysics*, 324, 111–134.

Berthelsen, A. (1953). On the geology of the Rupshu district, NW Himalaya. *Medd. Fra. Dansk. Geol.* Forening, Kobenkhavn, 12, 350–415.

Burg, J.P., Nievergelt, P., Oberli, F., Seward, D., Davy, P., Maurin, J.C., Diao, Z., Meier, M. (1998). The Namche Barwa syntaxis: Evidence for exhumation related to compressional crustal folding. *Journal of Earth Sciences*, 16, 239–252.

Chakungal, J., Dostal, J., Grujic, D., Duchêne, S., Ghalley, S.K. (2010). Provenance of the Greater Himalayan Sequence: Evidence from mafic eclogite-granulites and amphibolites in NW Bhutan. *Tectonophysics*, 480, 198–212, doi:10.1016/j.tecto.2009.10.014.

Chatterjee, N. and Ghose, N.C. (2010). Metamorphic evolution of the Naga Hills eclogite and blueschist, Northeast India: Implications for early subduction of the Indian plate under the Burma microplate. *Journal of Metamorphic Geology*, 28, 209–225.

Chen, Y., Chen, S., Guillot, S., Li, Q. (2022). Changes in subduction dip angle of the Indian Plate inferred from the western Himalaya eclogites. *Frontiers in Earth Science*, 9, 790999.

Chopin, C. (1984). Coesite and pure pyrope in high-grade blueschists of the Western Alps. *Contributions to Mineralogy and Petrology*, 86, 107–118.

Corrie, S.L., Koh, M.J., Verrvoort, J.D. (2016). Young eclogite from the Greater Himalayan Sequence, Arun Valley, eastern Nepal: P–T–t path and tectonic implications. *Earth and Planetary Science Letters*, 289, 406–416.

Cottle, J.M., Jessup, M.J., Newell, D.L., Horstwood, M.S.A., Noble, S.R., Parrish, R.R. (2009). Geochronology of granulitized eclogite from the Ama Drime Massif: Implications for the tectonic evolution of the South Tibetan Himalaya. *Tectonics*, doi: 10.1029/2008TC002256.

Ding, L., Zhong, D.L., Yin, A., Kapp, P., Harrisson, T.M. (2001). Cenezoic structural and metamorphic evolution of the eastern Himalayan syntaxis (Namche Barwa). *Earth and Planet Science Letters*, 192, 423–438.

Ernst, W.G. (1973). Blueschist metamorphism and P-T regimes in active subduction zones. *Tectonophysics*, 17, 255–272.

Franck, W., Gansser, A., Trommsdorf, K. (1977). Geological observations in the Ladakh area (Himalayas): A preliminary report. *Schweitz und Mineralogy and Petrography Mittleigung*, 57, 89–113.

Groppo, C., Lombardo, B., Rolfo, F., Pertusati, P. (2007). Clockwise exhumation path of granulitized eclogites from the Ama Drime rage (Eastern Himalayas). *Journal of Metamorphic Geology*, 25, 51–75.

Groppo, C., Rolfo, F., Sachan, H.K., Rai, S.K. (2016). Petrology of blueschist from the Western Himalaya (Ladakh, NW India): Exploring the complex behavior of a lawsonite-bearing system in a paleo-accretionary setting. *Lithos*, 252, 41–56.

Guillot, S., de Sigoyer, J., Lardeaux, J.M., Mascle, G. (1997). Eclogitic metasediments from the Tso Morari area (Ladakh, Himalaya): Evidence for continental subduction during India–Asia convergence. *Contribution to Mineralogy and Petrology*, 128, 197–212.

Guillot, S., Hattori, K.H., de Sigoyer, J., Nägler, T., Auzende, A.L. (2001). Evidence of hydration of the mantle wedge and its role in the exhumation of eclogites. *Earth and Planetary Sciences Letters*, 193(1–2), 115–127.

Guillot, S., Mahéo, G., de Sigoyer, J., Hattori, K.H., Pêcher, A. (2008). Tethyan and Indian subduction viewed from the Himalayan high- to ultrahigh-pressure metamorphic rocks. *Tectonophysics*, 451, 225–241.

Guillot, S., Hattori, K., Agard, P., Schwartz, S., Vidal, O. (2009). Exhumation processes in oceanic and continental subduction contexts: A review. In *Subduction Zone Dynamics?*, Lallemand, S. and Funiciello, F. (eds). Springer-Verlag, Berlin, Heidelberg. doi 10.1007/978-3-540-87974-9.

Hetényi, G., Cattin, R., Brunet, F., Bollinger, L., Vergne, J., Nábělek, J.L., Diament, M. (2007). Density distribution of the India plate beneath the Tibetan Plateau: Geophysical and petrological constraints on the kinetics of lower-crustal eclogitization. *Earth and Planetary Science Letters*, 264, 226–244.

Honegger, K., Le Fort, P., Mascle, G., Zimmerman, J.L. (1989). The blueschists along the Indus Suture Zone in Ladakh, NW Himalaya. *Journal of Metamorphic Geology*, 7, 57–72.

Iqbal, M.Z., Li, W., Ali, A., Liu, Y., Zhang, D. (2020). Clockwise hairpin-type metamorphic pressure–temperature (P–T) path recorded in the Shangla blueschist along the Indus Suture Zone, Pakistan Himalaya. *Geological Journal*, doi: 10.1002/gj.3878.

Kaneko, Y., Katayama, I., Yamamoto, H., Misawa, K., Ishikawa, M., Rehman, H.U., Kausar, A.B., Shiraishi, K. (2003). Timing of Himalayan ultrahigh-pressure metamorphism: Sinking rate and subduction angle of the Indian continental crust beneath Asia. *Journal of Metamorphic Geology*, 21, 589–599.

Kellett, D.A., Cottle, J.M., Smit, M. (2014). Eocene deep crust at Ama Drime, Tibet: Early evolution of the Himalayan orogen. *Lithosphere*, 6, 220–229.

Lanari, P., Riel, N., Guillot, S., Vidal, O., Schwartz, S., Pêcher, A., et al. (2013). Deciphering high-pressure metamorphism in collisional context using microprobe mapping methods: Application to the Stak Eclogitic Massif (Northwest Himalaya). *Geology*, 41(2), 111–114. doi:10.1130/G33523.1.

Le Fort, P., Guillot, S., Pêcher, A. (1997). HP metamorphic belt along the Indus suture zone of NW Himalaya: New discoveries and significance. *Compte Rendus de l'Académie des Sciences, Paris*, 325, 773–778.

Leech, M.L., Singh, S., Jain, A.K., Klemperer, S.L., Manickavasagam, R.M. (2005). The onset of India-Asia continental collision: Early, steep subduction required by timing of UHP metamorphism in the western Himalaya. *Earth and Planet Science Letters*, 234, 83–97.

Liu, Y. and Zhong, D.L. (1997). Petrology of high-pressure granulites from the eastern Himalaya syntaxis. *Journal of Metamorphic Geology*, 15, 451–466.

Lombardo, B. and Rolfo, F. (2000). Two contrasting eclogite types in the Himalayas: Implications fror the Himalayan orogeny. *Journal of Geodynamics*, 30, 37–60.

Lombardo, B., Pertusati, P., Rolfo, F., Visonà, D. (1998). First report of eclogites from the E Himalaya: Implications for the Himalayan orogeny. *Memorie di Scienze Geologiche*, 50, 67–68.

Mahéo, G., Bertrand, H., Guillot, S., Keller, F., Capiez, P. (2004). The South Ladakh ophiolite (NW Himalaya), a crustal and upper mantle section of the same immature arc: Implications for the closure of the Neothethys. *Chemical Geology*, 203, 273–303.

Mahéo, G., Fayoux, C., Guillot, S., Garzanti, E., Capiez, P., Mascle, G. (2006). Geochemistry of ophiolitic rocks and blueschists from the Sapi-Shergol mélange (Ladakh, NW Himalaya, India): Implication for the timing of the closure of the Neo-Tethys ocean. *Journal of Asian Earth Sciences*, 26, 695–707.

Mukheerjee, B.K. and Sachan, H.K. (2001). Discovery of coesite from Indian Himalaya: A record of ultrahigh pressure metamorphism in Indian continental crust. *Current Science*, 81, 1358–1361.

Mukheerjee, B.K., Sachan, H.K., Ogasawaray, Y., Muko, A., Yoshioka, N. (2003). Carbonate-bearing UHPM rocks from the Tso-Morari Region, Ladakh, India: Petrological implications. *International Geology Review*, 45, 49–69.

Nábělek, J., Hetényi, G., Vergne, J., Sapkota, S., Kafle, B., Jiang, M., Su, H. (2009). Underplating in the Himalaya-Tibet collision zone revealed by the Hi-CLIMB experiment. *Science*, 325(5946), 1371–1374.

O'Brien, P. (2019). Eclogites and other high-pressure rocks in the Himalaya: A review. *Geological Society, London, Special Publications*, 483(1), 183–213.

O'Brien, P., Zotov, N., Law, R., Khan, A.M. (2001). Coesite in Himalaya eclogite and implications for models of India–Asia collision. *Geology*, 29, 435–438.

Petley-Ragan, A., Ben-Zion, Y., Austrheim, H., Ildefonse, B., Renard, F. (2019). Dynamic earthquake rupture in the lower crust. *Science Advances*, 5(7), eaaw0913.

Pognante, U. and Spencer, D.A. (1991). First record of eclogites from the High Himalayan belt, Kaghan valley (northern Pakistan). *European Journal of Mineralogy*, 3(3), 613–618.

Rajkumar, A. (2015). Prograde Histories in High-P to Ultra-high-P Metamorphic Rocks from Tibet and Northern India. PhD Thesis, University of Sydney.

Rehman, H.U., Lee, H.-Y., Chung, S.-L., Khan, T., O'Brien, P.J., Yamamoto, H. (2016). Source and Mode of the Permian Panjal Trap Magmatism: Evidence from Zircon U-Pb and Hf isotopes and trace element data from the Himalayan ultrahigh-pressure rocks. *Lithos*, 260, 286–299. doi:10.1016/j.lithos.2016.06.001.

Sachan, H.K., Mukherjee, B.K., Ogasawara, Y., Maruyama, S., Ishida, H., Muko, A., Yoshioka, N. (2004). Discovery of coesite from Indus suture zone ISZ Ladakh India: Evidence for deep subduction. *European Journal of Mineralogy*, 16, 235–240.

Saint-Onge, M.R., Rayner, N., Palin, R.M., Searle, M.P., Waters, D.J. (2013). Integrated pressure-temperature-time constraints for the Tso Morari Dome (Northwest India): Implications for the burial and exhumation path of UHP units in the Western Himalaya. *Journal of Metamorphic Geology*, 31, 469–504. doi:10.1111/jmg.12030.

Shams, F.A. (1972). Glaucophane-bearing rocks from near Topsin, Swat. First record from Pakistan. *Pakistan Journal of Scientific Research*, 24, 343–345.

de Sigoyer, J. (1998). Mécanismes d'exhumation des roches de haute pression basse température, en contexte de convergence continentale (Tso Morari, NO Himalaya). PhD Thesis, University Claude Bernard, Lyon.

de Sigoyer, J., Guillot, S., Lardeaux, J.M., Mascle, G. (1997). Glaucophane-bearing eclogites in the Tso Morari dome (eastern Ladakh, NW Himalaya). *European Journal of Mineralogy*, 9, 1073–1083.

de Sigoyer, J., Chavagnac, V., Blichert-Toft, J., Villa, I.M., Luais, B., Guillot, S., Cosca, M., Mascle, G. (2000). Dating the Indian continental subduction and collisional thickening in the northwest Himalaya: Multichronology of the Tso Morari eclogites. *Geology*, 28, 487–490.

de Sigoyer, J., Guillot, S., Dick, P. (2004). Exhumation processes of the high-pressure low-temperature Tso Morari dome in a convergent context (eastern-Ladakh, NW-Himalaya). *Tectonics*, 23(3), TC3003. doi: 10.1029/2002TC001492.

Steck, A., Spring, L., Vannay, J.C., Masson, H., Stutz, E., Bucher, H., Marchant, R., Tieche, J.C. (1993). Geological transect across the northwestern Himalaya in eastern Ladakh and Lahul (a model for the continental collision of India and Asia). *Eclogae Geologicae Helveticae*, 86(1), 219–263.

Wang, Y.-H., Zhang, L.-F., Li, S.-Z., Somerville, I. (2017). Zircon U–Pb dating and phase equilibria modelling of gneisses from Dinggye area, Ama Drime Massif, central Himalaya. *Geological Journal*. doi:10.1002/gj.3027.

Wilke, F.D.H., O'Brien, P.J., Gerdes, A. (2010). The multistage exhumation history of the Kaghan Valley UHP series, NW Himalaya, Pakistan, from U-Pb and 40Ar/39Ar ages. *European Journal of Mineralogy*, 22, 703–719. doi:10.1127/0935-1221/2010/0022-2051.

Wilke, F.D.H., O'Brien, P.J., Schmidt, A., Ziemann, M.A. (2015). Subduction, peak and multistage exhumation metamorphism: Traces from one coesite-bearing eclogite. *Lithos*, 231, 77–91. doi:10.1016/j.lithos.2015.06.007.

Zhang, Z., Xiang, H., Dong, X., Ding, H., He, Z. (2015). Long-lived high-temperature granulite-facies metamorphism in the Eastern Himalayan orogen. South Tibet. *Lithos*, 212–215, 1–15.

5

The Greater Himalayan Sequence – Tectonic, Petrographic and Kinematic Evolution of the Metamorphic Core Zone of the Himalayan Orogeny

Martin ROBYR[1], Rodolfo CAROSI[2], Salvatore IACCARINO[2]
and Chiara MONTOMOLI[2]

[1] *University of Lausanne, Switzerland*
[2] *University of Turin, Italy*

5.1. Introduction

The continental collision between India and Asia in the late Paleocene–early Eocene, about 59–50 Ma ago (e.g. Hodges 2000; Green et al. 2008, see Volume 1 – Chapter 1), generated the largest zone of active crustal deformation on Earth, the Himalaya. Since the onset of this collision, a significant amount of crustal shortening has been accommodated by folding and by the displacement of units of the Indian continental Plate along major faults and shear zones. These major N-dipping structures segment today the

Himalayan range into several parallel and fairly continuous tectonic units over more than 2,500 km along strike (Figure 5.1 and Volume 1 – Chapter 3). The kinematic evolution of the Himalayan range is largely controlled by the successive subduction and extrusion of large slices of the Indian crust which have been progressively accreted southward in the frontal part of the belt (Figure 5.1).

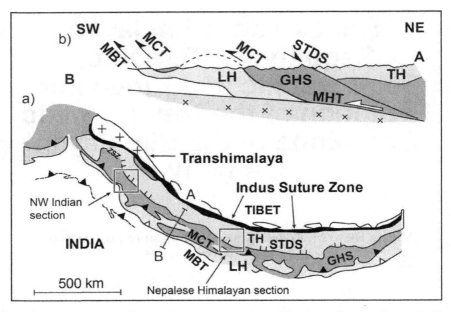

Figure 5.1. *(a) Generalized geological map of the Himalaya showing the cylindrical geometry of the orogeny. (b) Synthetic cross-section of the central Himalaya (Vannay and Grasemann 2001). HHC = High Himalayan Crystalline, THS = Tethyan Himalaya, LH = Lesser Himalaya, MBT = Main Boundary Thrust, MCT = Main Central Thrust, MHT = Main Himalayan Thrust, STDS = South Tibetan Detachment System*

One of these units, the Greater Himalayan Sequence (GHS, often also referred to as High Himalayan Crystalline, HHC), forms the crystalline core of the Himalayan range and corresponds to the geographical area with the highest reliefs. Throughout the Himalayan range, the GHS consists of a thick sequence of metasediments. These metasediments are the results of the metamorphism of original graywacke, siltite and pelite from the Haimantas Formation, deposited onto the passive margin of the Indian plate during the Precambrian to Cambrian time. Calcsilicate and marble, together with orthogneiss derived from Cambro-Ordovician intrusions, are also variably present.

The GHS gathers the rocks that have experienced the most intense regional metamorphism of the Himalayan range. During the collision, subduction processes and the ensuing crustal thickening generated intense regional deformation and metamorphism in this unit, including partial melting of the Precambrian sediments resulting in the generation of migmatite and leucogranitic bodies, commonly referred to as the High Himalayan Leucogranite (HHL). In most sections, all along the Himalayan range, this metamorphic core forms a north-dipping crustal slab bounded on its northern flank by the extensional structures of the South Tibetan Detachment System (STDS) and by a north-dipping thrust zone, referred to as the Main Central Thrust (MCT) along its southern flank (Figure 5.1). It is worth emphasizing that although the GHS is generally separated from the low-grade sedimentary series of the Tethyan Himalaya Sequence (THS) to the north by the extensional structures of the STDS, a gradual transition between the low-grade Haimantas and the high-grade paragneisses of the GHS is observed in several parts of the NW Himalaya. Numerous arguments suggest that the GHS paragneiss represents the metamorphic equivalents of the Haimantas, and not the true basement onto which the Tethyan Himalaya sediments were deposited.

The GHS records different metamorphic events. Prograded-path-related structures developed during the collision (Eocene–Early Oligocene) are overprinted by structures related to the exhumation stage (Oligocene–Miocene). The first event, related to prograde burial in the kyanite zone, is classically referred to as the Eo-Himalayan stage, followed by a second event, called the Neo-Himalayan stage, corresponding to a "nearly" isothermal decompression, down to andalusite zone, and by later cooling. It is commonly believed that the main exhumation of the GHS occurred during the later stage.

GHS metamorphism shows several important and peculiar features. The most famous one, recognized since the late 19th century, is perhaps the occurrence of an inverted metamorphic sequence (IMS) at the base of GHS. Indeed, moving structurally upwards, from the MCT to the top of the unit, this inverted metamorphic zonation is usually characterized by a gradual superposition of garnet, staurolite, kyanite, sillimanite + muscovite and sillimanite + K-feldspar isograds (Figure 5.2).

Despite over a century of geological investigations, and several tectono-metamorphic models proposed (including numerical ones), the developement of the IMS is still a debated topic within the geological community. The parallelism between the inverted isogrades and the structures of the MCT at regional scale suggests a close relation between the inverted metamorphism sequence (IMS) observed at the front and the tectonic activity along the MCT. Another puzzling feature of the GHS metamorphism is the identification, along several structural transects, of two zones characterized by very high "apparent" thermal field gradients, both at the lower and upper GHS limits. Interestingly, these two zones, which suggest the condensation of paleo-isotherms/surfaces, are associated with first-order regional scale ductile shear zones, the aforementioned MCT at the bottom and the STDS at the top, pointing out a fundamental link between the final pattern of metamorphism and deformation.

The GHS is characterized by a single, monoclinal slab plunging towards the north. This rather simple geometry is observed over a large part of the Himalayan range, in particular in its central and eastern parts. Recent multidisciplinary studies revealed that the geometry of the GHS is actually not as cylindrical as commonly thought. In the Nepalese Himalaya and in the NW Indian Himalaya, an architecture of the GHS developing a much more complex internal structure has been recognized.

In the Nepalese Himalaya, the GHS can be split into (at least) two different tectono-metamorphic units, with contrasting and diachronic metamorphic and deformative history, referred to as the lower GHS (GHSL) and the upper GHS (GHSU) (e.g. Corrie and Kohn 2011; Montomoli et al. 2013; Carosi et al. 2019).

The tectonic boundary between the two portions is represented by a contractional, high-temperature ductile shear zone, commonly referred to as the High Himalayan Discontinuity (HHD) active from the late-Oligocene to Miocene (i.e., 28–17 Ma) (Montomoli et al. 2013; Carosi et al. 2018, 2019). Similarly, in the NW part of India, the geologic structures and metamorphic zonations differ significantly from the classical NE-dipping stacking of the Himalayan unit that characterized the Himalayan architecture in most of the Himalayan section (Steck et al. 1993; Robyr and Lanari 2020, and references therein). A major feature of the tectono-metamorphic evolution of the GHS in NW India is that the main phase of metamorphism and tectonism relates to NE-directed thrusting. This is clearly in contrast with the southward thrusting and folding that has been predominant in the Himalaya since continental collision.

The Greater Himalayan Sequence 109

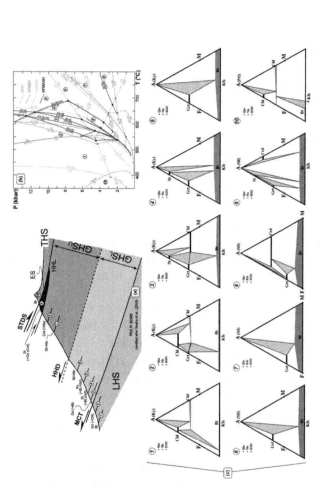

Figure 5.2. (a) Simplified cross-section of the Himalayan metamorphic core, from the LHS up to the THS. The tectono-metamorphic architecture, with a schematic distribution of metamorphic zones (numbers) based on the index-mineral distribution, is emphasized (modified after Searle et al. 2008). (b) Simplified petrogenetic grid for metapelite. Possible and indicative P–T conditions of the various metamorphic zones are indicated. (c) Serie of AFM diagrams, showing the sequence of metamorphic mineral assemblages, commonly observed in the metapelite from the metamorphic core. Numbers, referring to the possible metamorphic zones, are the same as of Figures 5.2(a) and (b). Figures 5.2(b) and (c) are based on Spear (1993, 1999). Mineral abbreviations after Kretz (1983). For a color version of this figure, see www.iste.co.uk/cattin/himalaya2.zip

In this chapter, the main characteristics of the deformation and metamorphism recorded in the rocks forming the GHS are presented along two reference sections located in the central part of the Himalaya of Nepal and the northwestern part of India.

5.2. Tectono-metamorphic evolution of the GHS in the central part of the Himalaya in Nepal

The Nepalese Himalaya has been a key area to define the structural and metamorphic settings of the Himalaya orogeny (e.g. Colchen et al. 1986; Vannay and Hodges 1996). This section presents a brief overview of the GHS features in Central Himalaya with a focus on a complete structural section of the GHS cropping out in the Mid-West Nepal along the Mugu–Karnali valley (Figure 5.3) (Montomoli et al. 2013; Iaccarino et al. 2017). This section will serve as a reference section for the GHS metamorphism. The geological setting and the main tectono-metamorphic units including their tectonic boundaries, lithologies and index-mineral distribution are illustrated by a simplified geological map and of a litho-tectonic column in Figures 5.3 and 5.4. A more comprehensive compilation is given in the detailed reviews of Guillot (1999); Kohn (2014); Goscombe et al. (2018) and Waters (2019).

The Himalayan belt (Figure 5.1(a)) is mainly composed of continuous packages of imbricated litho-tectonic units (Le Fort 1975; Hodges 2000). Among these units, the GHS (Figure 5.1), made of medium- to high-grade metamorphic rocks and Cenozoic leucogranites (HHL, Visonà et al. 2012), represents the exhumed mid-crustal core of the Himalaya (Hodges 2000; Cottle et al. 2015). Classically, in Nepalese Himalaya, the GHS has been subdivided into three main units (e.g. Vannay and Hodges 1996; Searle et al. 2008). The structurally lower unit 1 consists of garnet- and kyanite-bearing metapelites, subordinate quartzites, calcsilicates, migmatites and marbles. The structurally intermediate unit 2 is composed of medium- to high-grade calcsilicates, minor marbles and metapelites. The structurally highest unit 3 consists of orthogneiss and kyanite/sillimanite-bearing migmatites, with minor calcsilicate gneiss. Although it is widely used and suitable for several parts of the belt (e.g. Vannay and Hodges 1996), this lithological subdivision shows a strong heterogeneity along the strike of the belt, including the Mugu–Karnali area.

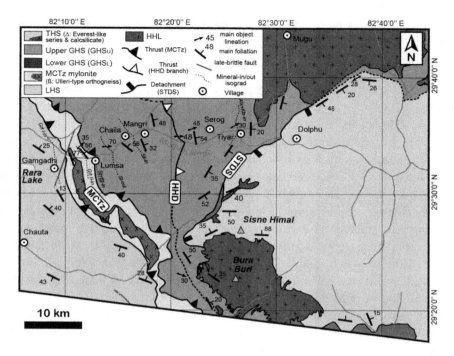

Figure 5.3. *Simplified geological map of the Mugu–Karnali area (Mid-Western Nepal) and surroundings (modified after Iaccarino et al. 2017; Carosi et al. 2019). For a color version of this figure, see www.iste.co.uk/cattin/himalaya2.zip*

Two main peculiarities of the GHS have to be taken into account (Hodges 2000, and references therein):

i) The GHS is tectonically bounded by two crustal scale shear zones with opposite kinematics (Figure 5.2); at the base of the GHS, the MCT is a ductile-to-brittle reverse shear zone that places the GHS above the LHS. Conversely, the upper tectonic contact of the GHS corresponds to the STDS, a system of ductile detachment and brittle normal faults. These NE-dipping extensional structures tectonically juxtaposed the low-grade sedimentary rocks of the THS over the high-grade GHS rocks. The GHS is thereby a medium–high grade tectono-metamorphic unit tectonically sandwiched between two units (LHS and THS), showing lower metamorphic conditions (Figure 5.2).

ii) The identification of an IMS at the GHS base, where high-grade "Barrovian" index-minerals appear progressively (from garnet to sillimanite) moving structurally upward (Figure 5.2). This feature is spatially, and likely genetically, related to the MCT (e.g. Vannay and Grasemann 2001; Searle et al. 2008).

As anticipated above, the metamorphic history in the GHS is classically subdivided into two main events (e.g. Vannay and Hodges 1996; Hodges 2000): (1) the Eo-Himalayan stage, which occurs during the Eocene–Oligocene, is responsible for the formation of the high-pressure prograde kyanite-bearing assemblages (Vannay and Hodges 1996); (2) the Neo-Himalayan event corresponds to the main period of exhumation initiated during Early Miocene. This latter event is characterized by coeval movement along both the STDS and MCT. During this phase, decompression leads to the development of medium- to low-pressure (sillimanite- or even cordierite-) bearing assemblages and extensive partial melting (and HHL production) within the GHS (e.g. Godin et al. 2006, and references therein). Such a subdivision into two distinct metamorphic events, despite very convenient and still useful, has to be rather regarded as a continuum evolution (as also done in the Alpine belt, see e.g. De Graciansky et al. 2010). Moreover, recent petrochonological investigations reveal a more complex metamorphic and thermal structure for the different portions of the GHS. For instance, the upper portions of the GHS experienced decompression and cooling nearly contemporaneously with prograde burial in the structurally lower GHS parts (Carosi et al. 2018, 2019, for a review).

The Mugu–Karnali valley (Figures 5.1 and 5.3), trending nearly E–W, provides almost 35 km of continuous outcrops along the main tectono-metamorphic units of the belt. The area was partly mapped in the 1970s (Fuchs 1974, 1977, and references therein) and re-investigated recently (Montomoli et al. 2013; Iaccarino et al. 2017; Carosi et al. 2019). The general architecture at the first glance corresponds to a NE–E-dipping monoclinal slab. The lower tectonic unit, the LHS, is composed of quartzite often showing primary structures, marble, dolomitic marble, graphitic schist and minor amphibolite. The main foliation in LHS is parallel to the axial plane of tight to isoclinal folds; it strikes NW–SE and dips to NE (Figure 5.3). Meso- and micro-structural analyses reveal that the main foliation in LHS is associated with a second deformation phase. This is testified by relicts of an older continuous to spaced foliation preserved in microlithons and as inclusion in porphyroblasts. The object lineation in LHS generally trends NE–SW and moderately plunges to the NE (Figure 5.3). Deformation and metamorphism increase structurally upward, with the development of garnet–chlorite–biotite-bearing phyllite (Grt I zone) just below the tectonic contact between the LHS and the GHS (Figures 5.3 and 5.4). These phyllites equilibrated at "peak metamorphic" conditions of 560°C and 7–8 kbar (Iaccarino et al. 2017).

The tectonic contact between GHS and LHS, in the studied area, consists of km-thick ductile shear zone affecting both quartzite and garnet–biotite-bearing phyllites (Grt II zone) and higher-grade metamorphic rocks of the GHS (Figures 5.3 and 5.4). For this reason, this shear zone is here referred to as the Main Central Thrust Zone (MCTz in Figures 5.3 and 5.4) which likely encompass a possible protolith boundary. Within the MCTz, an NW–SE striking mylonitic foliation is developed. Kinematic indicators such as S–C–C' fabric, asymmetric porphyroclasts and quartz sigmoids point to a top-to-the-WSW sense of shear. Applying the structurally based definition of Searle et al. (2008), the MCTz should be placed wholly within the GHS; in contrast, based on lithostratigraphic arguments, DeCelles et al. (2020) mapped the base of the GHS at the vicinity of the top of the MCTz mylonitic belt in the study area (Figure 5.4).

Moving upsection, the main GHS foliation, dipping mainly towards E, strikes from NW–SE to NNE–SSW. An earlier foliation is sporadically observed at both the meso- and micro-scales. The object lineation in GHS trends mainly NE–SW and plunges moderately to the E–NE. Moving structurally upward, an inverted metamorphic gradient within the GHS is characterized by metamorphic conditions increasing from garnet–biotite-bearing rocks up to garnet–kyanite–biotite paragneiss and sillimanite-bearing migmatitic gneiss (Figures 5.3, 5.4 and 5.5). In the Mugu–Karanali, the GHS has been subdivided into two portions. the lower GHS (GHSL) and the upper GHS (GHSU). These subunits are separated by a tectonic and metamorphic discontinuity, the Mangri Shear Zone described in detail by Montomoli et al. (2013) and Iaccarino et al. (2017). The Mangri Shear Zone is interpreted as the local branch on the HHD (Montomoli et al. 2013, 2015; Carosi et al. 2018, 2019).

The GHSL consists mainly of garnet-, staurolite-, to kyanite-bearing metapelite, quartzite and orthogneiss, with minor calcsilicate and amphibolite (Figures 5.3, 5.4 and 5.5). The first appearance of sillimanite, as rare fibrolite, has been documented within orthogneiss, SW of the village of Mangri (Figures 5.3 and 5.4). P–T estimates of Montomoli et al. (2013) and Iaccarino et al.

(2017) suggest that the GHS∟ experienced clockwise P–T paths, with P–T conditions of 9–12 kbar and 600–700°C.

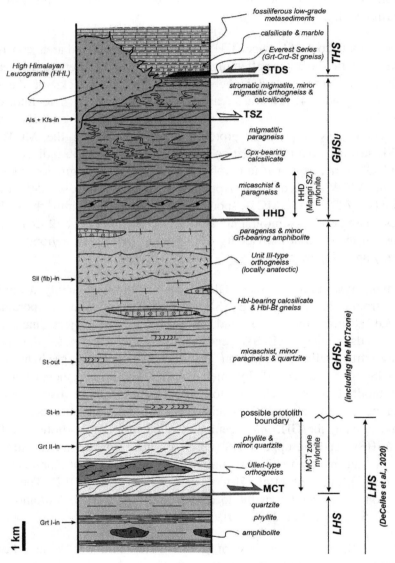

Figure 5.4. *Litho-tectonic column summarizing the geological settings of the Mugu–Karnali structural transect. The main tectono-metamorphic units and their tectonic boundaries, lithologies and index-mineral distribution are given (based on data of Iaccarino et al. 2017; DeCelles et al. 2020). For a color version of this figure, see www.iste.co.uk/cattin/himalaya2.zip*

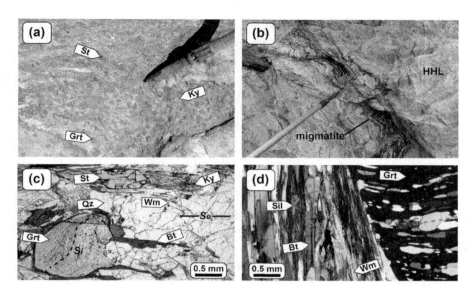

Figure 5.5. *Example of metamorphic rocks, at the outcrop- and micro-scale, along the Mugu–Karnali transect. (a) Barrovian porphyroblasts in a paragneiss of the GHSL. (b) Migmatitic gneiss of GHSU partially enveloped within a decametric body of HHL. (c) Barrovian minerals at the micro-scale from the GHSL. Note the occurrence of an internal foliation (S_i, defined by opaque-minerals) inside garnet. (d) Barrovian minerals at the micro-scale. Note the internal foliation, S_i, (defined by quartz) inside garnet that is not in-continuity with the external one, S_e, defined by fibrolitic sillimanite and micas (GHSU). For a color version of this figure, see www.iste.co.uk/cattin/himalaya2.zip*

Structurally upward, sillimanite and white mica-bearing mylonitic paragneiss and micaschist of the Mangri Shear Zone mark the bottom of the GHSU (Figures 5.3, 5.4 and 5.5). Mylonitic foliation of the Mangri Shear Zone is concordant with the main foliation in the host GHS wall-rocks. It strikes NW–SE and dips moderately (20–50°) towards the NE. Object lineation of sillimanite- and quartz rods trends NW–SE to nearly W–E and plunges moderately (30–40°) to the SE/E. Kinematic indicators at the meso- and micro-scales, such as S–C–C' fabric, rotated σ-type porphyroclasts (mainly garnet and feldspar) and mica-fish point to a top-to-the-S/SW sense of shear. Petrographic and microstructural arguments suggest that shearing along the Mangri Shear Zone was active at upper-amphibolite facies conditions and related to decompression from P–T conditions of 10–12 kbar at c. 650°C down to 7–8 kbar and c. 680–720°C (Montomoli et al. 2013; Iaccarino et al. 2017).

The GHSU contains sillimanite and white mica paragneiss and micaschist (Figure 5.5), passing upward to the migmatitic complex where K-feldspar + aluminosilicate + garnet-bearing migmatite (stromatic metatexite and minor diatexite), migmatitic orthogneiss and minor clinopyroxene-bearing calcsilicate are common (Figures 5.3, 5.4 and 5.5).

The base of the migmatitic complex is marked by a high-temperature (HT) contractional shear zone, the Tiyar Shear Zone (TSZ) of Iaccarino et al. (2017) (Figures 5.3 and 5.4) that coincides with the disappearance of white mica (the aluminosilicate + K-feldspar isograd). The main mylonitic foliation of the TSZ, reworking the migmatitic fabric, is concordant with the foliation observed in the host gneisses. The object lineation, dipping to the E–NE, is marked by sillimanite and quartzofeldspathic minerals. Top-to-the-W–SW kinematic indicators, such as quartzofeldspathic sigmoidal pods (Figure 5.2(d)), S–C fabric, garnet, kyanite and K-feldspar porphyroclasts, are present. In the GHSU, including the mylonites of the Mangri Shear Zone, the occurrence of kyanite, variably replaced by sillimanite, reveals that the GHSU also experienced a previous HP stage. Metamorphic data of Iaccarino et al. (2017) indicate that the migmatitic complex reached peak conditions, close to the HP-granulite facies (750°C, > 10 kbar), followed by nearly isothermal decompression down to 8–7 kbar and by later cooling.

The GHSU is intruded by HHL, forming numerous dykes, sills and up to large plutons such as the Mugu and Bura Buri leucogranites (Figures 5.3 and 5.4; e.g. Le Fort and France-Lanord 1995; Carosi et al. 2013). The Mugu leucogranite, with an intrusion age of 17.6 ± 0.3 Ma (Th–Pb on monazite, Harrison et al. 1999), contains K-feldspar–plagioclase–quartz–white mica, minor biotite and variable amounts of peraluminous accessory phases such as tourmaline, garnet and sillimanite.

South of the Mugu granite (Figure 5.3), structurally above the migmatitic complex, leucogranitic dyke swarms intrude intercalations of calcsilicate and low-grade biotite-bearing metapsammopelite, passing upward to low-grade "lumachelle"-bearing metalimestone, interpreted as the base of the THS (Figures 5.3 and 5.4). The main foliation strikes ENE–WSW dipping towards S–SE. Biotite lineations present an ENE–WSW trend and plunge to the

SW. The base of these rocks is strongly deformed with zoned calcsilicates showing both symmetric and asymmetric boudins. Rare kinematic indicators such as flanking folds point to a top-to-the-S/SE sense of shear. The observation of non-coaxial ductile shearing, kinematics indicators and of a normal-right-way-up metamorphic gradient (from c. 600°C down to <400°C) are linked to the effect of the ductile STDS. However, the inaccessibility of some areas surrounding the Mugu granite does not allow us to map this region entirely and to robustly constrain the relationships between the granite, the STDS and the THS. This relation can nevertheless be evaluated by comparing it with what is observed further south, around the Bura Buri granite (Figure 5.3; Carosi et al. 2013). Here, the surroundings of the Bura Buri granite consist of garnet–cordierite (after staurolite)-two mica-bearing gneiss in correlation with the Everest Series of Jessup et al. (2008), calcsilicate marble to low-grade marble and "lumachelle" limestone of the THS.

The whole tectonic pile, including the LHS and the entire GHS, is heterogeneously affected by a late ductile deformation event. This is marked by upright folds often with kink-like geometry, likely linked to the large-scale nappe re-folding (forming the Dolpo syncline and anticline, Fuchs 1977). No syn-kinematic mineral recrystallization on the axial plane-related foliation of the late folds is developed, suggesting that such a late event occurred at higher structural levels. The late folding event is followed by brittle tectonics, with different fault types and kinematics, that still deserve further studies.

5.3. Tectono-metamorphic evolution of the GHS in the north-western part of the Indian Himalaya in Himachal Pradesh and Ladakh

One of the characteristics of the Himalayan range is its apparent cylindrical geometry marked by the apparent remarkable lateral continuity of the different tectonic elements over the entire Himalayan range from the western to the eastern Himalayan syntaxis (Figure 5.1). The Himalayan mountain belt is thus classically subdivided into subparallel tectonics units, including the Lesser Himalaya (LH) at the front, the Greater Himalayan Sequence (GHS) that corresponds to the metamorphic core zone of the range and the Tethyan Himalaya (TH) formed by weakly metamorphosed sediment farther north.

For more than 1,400 km along the range, in the central and eastern parts of the Himalaya, the GHS corresponds to a fairly monoclinal, NE-dipping slab up to 20 km thick, mainly exposed in the frontal part of the orogeny (Figure 5.1). According to the classical view of the kinematic evolution of the GHS, the metamorphic imprint in this unit is interpreted as resulting from its northward underthrusting below the Tethyan Himalaya. During this middle Eocene to late Oligocene tectonic phase, the sediments forming now the high-grade rocks of the GHS were buried down to 40 km depth, where temperatures up to 850°C triggered amphibolite facies metamorphism and partial melting.

All along the belt, this GHS thrusts over the low- to medium-grade sedimentary series of the Lesser Himalaya along the Main Central Thrust; a major intracontinental thrust developed within the Indian margin during Early Miocene. In most Himalayan sections, the GHS is separated from the low-grade sedimentary series of the Tethyan Himalaya by the NE-dipping extensional structures of the South Tibetan Detachment System (STDS).

Several studies demonstrated that extension along the STDS and thrusting along the MCT initiated synchronously during Early Miocene at ~23 Ma (Frank et al. 1977; Hubbard and Harrison 1989). Yet, in central Nepal, the onset of the exhumation of the GHS took place before (middle–late Eocene) the period of MCT and STDS activities (Carosi et al. 2013; Iaccarino et al. 2017; Montemagni et al. 2020). Movements along both the MCT and the STDS nevertheless suggest a southward tectonically controlled extrusion of the high-grade paragneiss of the GHS. This relatively simple geometry reflects a kinematic evolution of the Himalayan range mainly controlled by the successive exhumation of large Indian crustal slices that have been detached from the underthrusting Indian plate and sequentially accreted towards the south along major NE-dipping faults and shear zones.

Yet, in the NW Indian Himalaya, between the Kullu valley and the downstream part of the Chenab valley, the hanging wall of the MCT, in the frontal part of the range, mainly consists of low-grade metasediments. The high-grade metamorphic rocks of the GHS are exposed farther north, as a large-scale dome structure called the Gianbul dome (Figure 5.6).

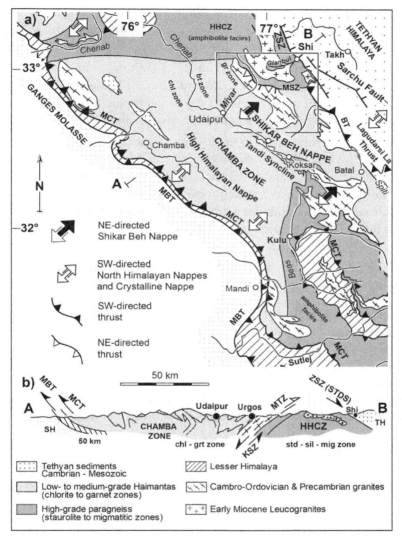

Figure 5.6. (a) Geological map of the NW Indian Himalaya (compiled after Steck et al. 1999; Vannay and Grasemann 2001). (b) General cross-section of the NW Indian Himalaya across the Gianbul dome (modified after Steck et al. 1999). HHC = High Himalayan Crystalline, NHC = North Himalayan Crystalline, TH = Tethyan Himalaya, LH = Lesser Himalaya, SH = Sub-Himalaya, MBT = Main Boundary Thrust, MCT = Main Central Thrust, STDS = South Tibetan Detachment System, ZSZ = Zanskar Shear Zone, MSZ = Miyar Shear Zone, BT = Baralacha La Thrust

The metamorphic record and geologic structures in the NW part of the Himalaya of India are thus significantly different. In this region, the extensional structure of the ZSZ and the thrust zone of the MCT encompass a tectonic unit that is not exclusively made of high-grade metamorphic rocks such as observed in the GHS of central and eastern Himalaya. In this part of the Himalayan range, the geometry of the metamorphic zonation and tectonic structure allows defining two metamorphic zones within the GHS (Figure 5.6):

– The Chamba zone refers to the low-grade metasediments exposed in the hanging wall of the MCT, in the frontal part of the High Himalayan Nappe. Towards the NE, this Chamba zone is cartographically connected to the basal part of the Tethyan Himalaya, which consist of a several kilometers thick series of Neo-Proterozoic to Cambrian detrital sediments (graywackes, siltstones and pelites), referred to as the Haimantas Formation.

– The High Himalayan Crystalline Zone of Zanskar (HHCZ) that is exposed in the core of the Gianbul dome and that corresponds to the metamorphic equivalent of the Haimantas (Robyr et al. 2002).

In addition, this region of the Himalayan range is distinguished also by the singular vergence of the tectonic structures.

Cross-sections along most of the Himalayan transects reveal that SW-vergent folding and thrusting have been the predominant tectonic structures developed within the Indian continental crust to accommodate crustal shortening since the onset of continental collision between India and Asia $\sim$55 Myr ago (Figure 5.1). However, over the last decades, NE-vergent structures have been described in the GHS of Upper Lahul in the northwestern Indian Himalaya. These structures consist of NE-vergent folds and NE-directed thrust faults, in sharp contrast to the SW-directed folding and thrusting that characterize much of the Himalayan kinematic since the continental collision.

One of these structures, a spectacular NE-directed thin-skinned thrust structure, referred to as the Lagudarsi La thrust, is exposed near the village of Kiato in the upstream part of the Spiti valley (Figure 5.7) (Steck et al. 1998). In this area, structural interference patterns reveal that the NE-directed structures associated with the Lagudarsi La thrust predate the main tectonic phase marked by SW-directed structures. Additional NE-directed structures have also been reported farther to the west along the Chamba River. In this region, the NE-verging Tandi syncline constitutes probably the most remarkable geologic feature supporting the fact that in the Upper Lahul region the earliest stage of Himalayan tectonism is related to NE-directed movements (Figures 5.6 and 5.7).

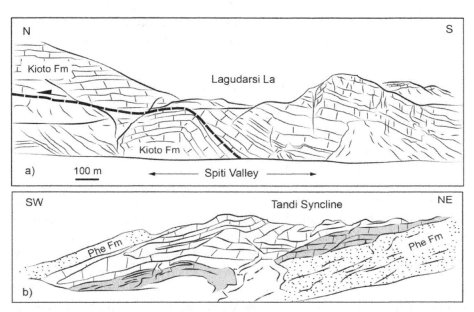

Figure 5.7. *Panoramic views of two major NE-directed tectonic structures observed in the Upper Lahul region: a) the NE-directed Lagudarsi La Thrust in the upper Spiti valley near the village of Kiato (from a drawing of Steck et al. 1998); b) panoramic view of the closure of the Tandi syncline towards the south (from a drawing of Steck et al. 1993)*

The Tandi syncline that corresponds to a synformal fold closing to the SW is exposed in the Pir Panjal range, south of Tandi. This structure consists of Permian and Mesozoic sediments folded into paragneiss of Upper Precambrian to Cambrian age (Phe Formation). A detailed description and structural analysis of the Tandi syncline by Vannay and Steck (1995) showed that the Tandi syncline corresponds to an NE-verging fold and that the formation of this syncline precedes the predominant Himalayan deformation manifested by folding towards the SW. Additional top to-the-NE structure has been observed within the Miyar Shear Zone that marks the transition between the HHCZ of Zanskar and the Chamba Zone on the southern limb of the Gianbul dome (Robyr et al. 2002).

The aim of the following sections is to describe the tectono-metamorphic evolution of the HHCZ based on the detailed structural, petrographic and geochronologic investigation along the Miyar Valley–Gianbul Valley transect (Figure 5.8) that represents a beautiful natural cross-section through the structures and metamorphic zonations in the core of the Gianbul dome.

5.3.1. Metamorphism and deformation in the High Himalayan Crystalline Zone of Zanskar

The Gianbul dome is cored by leucogranite and migmatitic paragneisses symmetrically surrounded by rocks of the sillimanite + K-feldspar, kyanite ± staurolite, garnet, biotite and chlorite zones (Figure 5.8).

On the northern limb of the dome, in the Gianbul valley, the contact between the high-grade rocks forming the core of the dome and the low-grade sediments of the Tethyan Himalaya corresponds to the NE-dipping Zanskar Shear Zone, a local equivalent of the STDS. This Early Miocene extensional structure initially acted as a thrust zone along which the rocks of the GHS, now forming the northern limb of the dome, were underthrust below the frontal part of the North Himalayan nappes that have affected the sedimentary series of the Tethyan Himalaya. Geochronological results from the northernmost part of the GHS of Zanskar established that the prograde metamorphism in this part of the range occurred between 35 and 25 Ma (Vance and Harris 1999). Within the Zanskar Shear Zone, the successive growing of sillimanite, cordierite, andalusite and margarite indicates a nearly isothermal retrograde P–T path suggesting that the rocks of the HHCZ of Zanskar were rapidly exhumed along the extensional structures of the Zanskar Shear Zone. U–Pb dating of monazite from various deformed and underformed leucogranitic dykes across the Zanskar Shear Zone indicates that the main extensional shearing along the Zanskar Shear Zone initiated shortly before 22.2 Ma and ceased by 19.8 Ma (Dèzes et al. 1999).

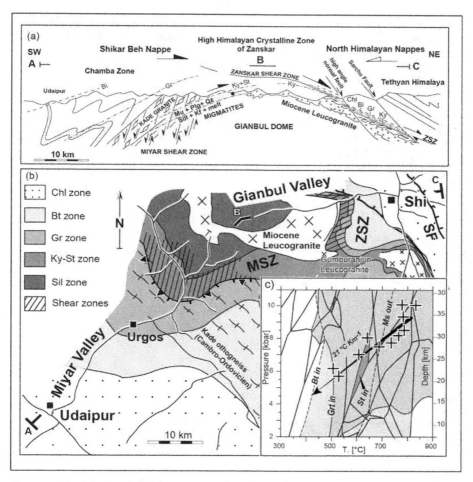

Figure 5.8. (a) Geological cross-section through the Gianbul dome along the Miyar valley and Gianbul valley (after Steck et al. 1999). Abbreviations: Chl = chlorite, Bi = biotite, Gr = garnet, St = staurolite, Ky = kyanite, Mu = muscovite, Plg = plagioclase, Qz = quartz, Sill = sillimanite, Kf = K-feldspar. (b) Metamorphic map of the Gianbul dome area (after Robyr et al., 2014). ZSZ = Zanskar Shear Zone, MTZ = Miyar Thrust Zone, KSZ = Khanjar Shear Zone. (c) Equilibrium phase diagram based on the bulk chemistry of a representative pelitic sample from the MSZ. The crosses correspond to P–T estimates for samples collected in the garnet, staurolite and kyanite and sillimanite zones after Robyr et al. (2002). The black arrow corresponds to the metamorphic field gradient inferred from the P–T estimates assuming a lithostatic gradient of 2.7 kbar/km (Steck 2003). Mineral abbreviations after Kretz (1983)

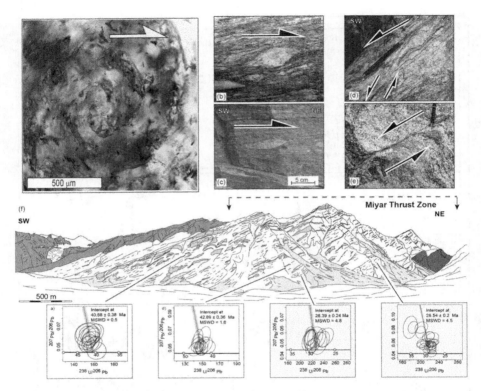

Figure 5.9. *(a–f) Representative structures from the Miyar Thrust Zone. (a) Sigmoidal inclusion trails in garnet from the Miyar Shear Zone. This synkinematic garnet indicates an NE-verging simple shear in the Miyar Thrust Zone zone. (b-c) Mylonitic amphibolite facies paragneisses and sandstones with sigma-type quartz porphyroclasts showing top-to-NE shear sense (lower part of the Miyar Thrust Zone). (d-e) Extensional shear bands of the normal Khanjar Shear Zone overprinting the main foliation in the migmatitic zone. (f) Panoramic view of the Miyar Shear Zone. Tera-Wasserburg plots for monazite from the different samples across the Miyar Shear Zone showing the age and the location of the analyzed monazite grains. For a color version of this figure, see www.iste.co.uk/cattin/himalaya2.zip*

To the south, the Gianbul dome is delimited along the Miyar valley by the SW-dipping Miyar Shear Zone (MSZ) (Pognante et al. 1990; Steck et al. 1999; Robyr et al. 2002). This structure marks the transition between the high-grade metamorphic rocks forming the core of the dome and the low- to medium-grade metasediments of the Chamba zone that forms in this region, the hanging wall of the MCT. This MSZ consists of an approximately 3 km wide SW-dipping shear zone containing a wide range of tectonic structures including sheath folds, sigma clasts, shear bands or back-rotated boudins that all show a top-to-the NE sense of shear (Figure 5.9). The development of sigmoidal inclusion trails in syntectonic garnets prophyroblasts relates the growth of the metamorphic assemblage with NE-directed tectonic movements (Figure 5.9).

Across the MSZ, the NE-verging contractional structures are superimposed by SW-dipping extensional shear bands and back-rotated boudins that indicate a top-to-the SW sense of shear (Figure 5.9). These observations indicate that the MSZ acted as an NE-directed synmetamorphic thrust before being reactivated as an SW-directed ductile zone of extension during the exhumation of the Gianbul dome.

The spiral-shaped geometry of inclusion trails preserved in garnet porphyroblasts in the footwall of the MSZ indicates simultaneous growth and rotation of the crystals during top-to-NE synmetamorphic deformation. Within the ZSZ, in contrast, sigmoidal inclusion trails indicate garnet crystallization during a top-to-SW synmetamorphic deformation. Consequently, the Barrovian metamorphic field gradient observed across the MSZ cannot be related to the NE-directed underthrusting of the GHS below the Tethyan Himalaya but must result from its underthrusting below the advancing front of a crustal thickening phase coming from the south (Robyr et al. 2014). Thermobarometric data from samples collected in the garnet and staurolite zone indicate a metamorphic field gradient of 21°C/km (Figure 5.8).

5.3.2. *Timing of crustal shortening and metamorphism along the Miyar Shear Zone*

In order to constrain the timing of the kinematics evolution of the MSZ and to test the hypothesis of an early Eocene nappe emplacement, samples were selected within the MSZ for monazite geochronology (Goswami-Banerjee and Robyr 2015; Robyr and Lanari 2020). Monazites dating yield ages of about 40 Ma for the structurally upper part of the shear zone, whereas monazite ages become significantly younger near the basal part of the shear zone, yielding ages ranging from 30 to 27 Ma. The geochronological data across the MSZ consequently show a sudden drop in the ages towards the north as the metamorphic conditions increase.

These data indicate that the Barrovian metamorphism along the Miyar Valley occurs in at least two successive stages, an Eocene phase and an Oligocene phase. So, the GHS in the Miyar valley seems to be constituted by two metamorphic slices that were successively underthrusted as a consequence of the NE-directed propagation of crustal deformation associated with the Shikar Beh nappe emplacement. In contrast, geochronological data from the northern part of the Gianbul dome indicate that the sediments forming now the high-grade rocks of the HHCZ on the northern flank of the dome were underthrusted below the frontal part of the North Himalayan nappes that have affected the sedimentary series of the Tethyan Himalaya. During this tectonic phase that occurred during the Oligocene, these rocks were buried down to 40 km depth, where temperature up to 850°C triggered amphibolite metamorphism and partial melting (Dèzes et al. 1999).

5.3.3. *Kinematic and tectonothermal evolution of the High Himalayan Crystalline Zone of Zanskar*

The structural, metamorphic and geochronological data indicate that the HHCZ in the Miyar valley consists of several tectonic slices that were successively southwestward underthrust as a consequence of NE-directed propagation of crustal deformation associated with the Shikar Beh nappe emplacement (Figure 5.10). An early phase of NE-directed crustal shortening was mainly accommodated by thrusting along the Salgaraon Thrust (Steck et al. 1999) in the Chenab valley southwest of Udaipur and by large scale NE-verging folding in the southern part of the Miyar valley. During this stage, the rocks situated nowadays in the hanging wall and the upper part of the MSZ were metamorphosed under P–T conditions prevailing for chlorite to garnet crystallization.

From the middle Eocene, a shear zone, that is, the Miyar Shear Zone, developed at the advancing front of this NE-directed folding phase, leading to the formation of the Shikar Beh nappe. During this stage, the metapelites of the Miyar valley in the footwall of the Miyar Shear Zone were buried to 25 km depth where metamorphic staurolite-bearing assemblages formed in metapelites (Robyr et al. 2002).

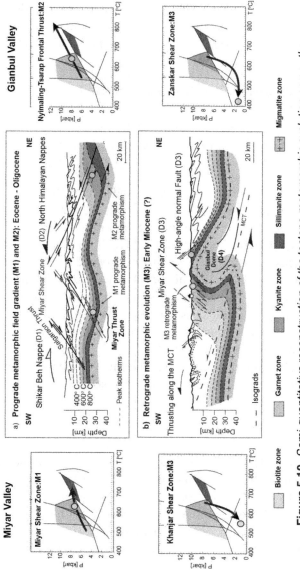

Figure 5.10. *Semi-quantitative reconstruction of the tectono-metamorphic evolution across the Gianbul dome, based on petrographic, geochronological and structural data. For a color version of this figure, see www.iste.co.uk/cattin/himalaya2.zip*

From this time on, the dynamics of the system weakened, probably because most of the shortening was since then accommodated by northward underthrusting of the HHCZ below the frontal part of the north Himalayan nappes between 33 and 28 Ma (e.g. Vance and Harris 1999). Beneath the Zanskar thrust, the rocks were subducted down to approximately 30 km depth, where temperatures up to 750–850°C triggered partial melting (Dèzes et al. 1999).

During the early Oligocene, ~28 Myr ago, the NE-directed thrust system at the front of the Shikar Beh nappe was reactivated as a ductile shear zone (Miyar Thrust Zone, MTZ), and rapid burial along the MTZ transforming the HHCZ rocks of the upstream part of the Miyar section into amphibolite-facies to migmatitic paragneiss.

During the late Oligocene (between 26.6 Ma and 23 Ma; Robyr et al. 2006), the onset of the exhumation of the Gianbul dome triggers the reactivation of the MSZ as a ductile zone of extension (Steck et al. 1999; Robyr et al. 2002). Simultaneously, the exhumation of the HHCZ as a dome structure was accompanied by extension along the ZSZ that reactivated the frontal thrust of the North Himalayan nappes (Patel et al. 1993). The rapid exhumation of the HHC high-grade rocks caused a nearly isothermal decompression resulting in partial migmatization of the paragneiss forming now the core of the Gianbul dome. Following an initial doming phase that ended by 22 Ma with the cessation of the movements along the MSZ (Robyr et al. 2006, 2014), further extension along the ZSZ and thrusting along the MCT lead to the tectonically extrusion of the High Himalayan nappe towards the south.

5.4. Conclusion

The GHS is classically seen as a wedge of the Indian continental crust formed by medium- to high-grade metamorphic rocks that has been uniformly extruded southward by coeval extensional movements along the STDS and by thrusting along the MCT (see Volume 1 – Chapter 3). Although this vision reflects fairly well the general kinematic evolution of the Himalayan range since the continental collision, studies clearly show that this vision is too simplistic.

Yet, it is factual that the GHS is, in most of sections, composed of high-grade metamorphic rocks derived from fairly homogeneous Precambrian to Cambrian detrital sediments over the entire Himalayan range. However, by looking at the detail of structural and petrographic data from various Himalayan sections, it appears that a range-parallel variability is indeed present. Some of the variabilities probably reflect local heterogeneities in strain during deformational phases, but most of these discrepancies likely result from differences in crustal-scale processes during the Himalayan orogeny.

Detailed regional studies in different areas of the Himalayan range, such as those presented in this chapter, reveal that the tectonic and metamorphic history of the Himalayan range, and in particular the kinematic evolution of the GHS, is not as cylindrical as the general tectonic maps of Himalaya lead us to believe (see also Volume 1 – Chapters 3, 5 and 6 for discussions about lateral variations).

5.5. References

Carosi, R., Montomoli, C., Rubatto, D., Visonà, D. (2013). Leucogranite intruding the South Tibetan Detachment in western Nepal. Implications for exhumation models in the Himalayas. *Terra Nova*, 25, 478–489.

Carosi, R., Montomoli, C., Iaccarino, S. (2018). 20 years of geological mapping of the metamorphic core across Central and Eastern Himalayas. *Earth-Science Reviews*, 177, 124–138.

Carosi, R., Montomoli, C., Iaccarino, S., Visonà, D. (2019). Structural evolution, metamorphism and melting in the Greater Himalayan Sequence in central-western Nepal. *Geological Society, London, Special Publications*, 483, 305–323.

Colchen, M., Le Fort, P., Pêcher, A. (1986). *Notice explicative de la carte géologique Annapurna-Manaslu-Ganesh (Himalaya du Népal) au 1:200.000eme (bilingue. Francais-English)*. CNRS, Paris.

Corrie, S.L. and Kohn, M.J. (2011). Metamorphic history of the Central Himalaya, Annapurna region, Nepal, and implication for tectonic models. *Geological Society of American Bulletin*, 123, 1863–1879.

Cottle, J.M., Larson, K.P., Kellett, D.A. (2015). How does the mid-crust accommodate deformation in large, hot collisional orogens? A review of recent research in the Himalayan orogen. *Journal of Structural Geology*, 78, 119–133.

De Graciansky, P-C., Roberts D.G., Tricart P. (2010). The Western Alps, from rift to passive margin to orogenic belt. An integrated geoscience overview. *Development in Earth Surface Processes*, 14, 429.

DeCelles, P.G., Carrapa, B., Ojha, P.T., Gehrels, G.E., Collins, D. (2020). Structural and thermal evolution of the Himalayan Thrust Belt in Midwestern Nepal, structural and thermal evolution of the Himalayan Thrust Belt in Midwestern Nepal. *Geological Society of America*, Special Paper, 547.

Dèzes, P.J., Vannay, J.C., Steck, A., Bussy, F., Cosca, M. (1999). Synorogenic extension. Quantitative constraints on the age and displacement of the Zanskar shear zone (northwest Himalaya). *Geological Society of America Bulletin*, 111, 364–374.

Frank, W., Thoni, M., Purtscheller, F. (1977). Geology and petrography of Kulu – South Lahul area. *Colloques Internationaux du Centre National de la Recherche Scientifique*, 268, 147–172.

Fuchs, G. (1974). On the geology of the Karnali and Dolpo Regions, West Nepal. *Mitteilungen der Geologischen Gesellschaft*, 66–67, 21–35.

Fuchs, G. (1977). The geology of the Karnali and Dolpo regions, Western Nepal, *Jahrbuch der Geologischen Bundesanstalt*, 120, 165–217.

Godin, L., Grujic, D., Law, R.D., Searle, M.P. (2006). Channel flow, ductile extrusion and exhumation in continental collision zones. An introduction. *Geological Society of London Special Publication*, 268, 1–23.

Goswami-Banerjee, S. and Robyr, M. (2015). Pressure and temperature conditions for crystallization of metamorphic allanite and monazite in metapelites. A case study from the Miyar Valley (high Himalayan Crystalline of Zanskar, NW India). *Journal of Metamorphic Geology*, 33, 535–556.

Goscombe, B., Gray, D., Foster, D.A. (2018). Metamorphic response to collision in the Central Himalayan Orogen. *Gondwana Research*, 57, 191–265.

Green, O.R., Searle, M.P., Corfield, R.I., Corfield, R.M. (2008). Cretaceous-tertiary carbonate platform evolution and the age of the India-Asia collision along the Ladakh Himalaya (northwest India). *Journal of Geology*, 116, 331–353.

Guillot, S. (1999). An overview of the metamorphic evolution in Central Nepal. *Journal of Asian Earth Sciences*, 17, 713–725.

Harrison, T.M., Grove, M., McKeegan, K.D., Coath, C.D., Lovera, O.M., Le Fort, P. (1999). Origin and episodic emplacement of the Manaslu Intrusive Complex, Central Himalaya. *Journal of Petrology*, 40, 3–19.

Hodges, K.V. (2000). Tectonics of the Himalaya and southern Tibet from two perspectives. *Geological Society of America Bulletin*, 112, 324–350.

Hubbard, M.S. and Harrison, T.M. (1989). $^{40}Ar/^{39}Ar$ age constraints on deformation and metamorphism in the Main Central Thrust zone and tibetan slab, eastern Nepal Himalaya. *Tectonics*, 8, 865–880.

Iaccarino, S., Montomoli, C., Carosi, R., Massonne, H., Visonà, D. (2017). Geology and tectono-metamorphic evolution of the Himalayan metamorphic core. Insights from the Mugu Karnali transect, Western Nepal (Central Himalaya). *Journal of Metamorphic Geology*, 35, 301–325.

Jessup, M.J., Cottle, M.J., Searle, M.P., Law, R.D., Newell, D.L., Tracy, R.J., Waters, D.J. (2008). P–T–t–D paths of Everest Series schist, Nepal. *Journal of Metamorphic Geology*, 26, 717–739.

Kohn, M.J. (2014). Himalayan metamorphism and its tectonic implications. *Annual Review of Earth and Planetary Sciences*, 42, 381–419.

Kretz, R. (1983). Symbols for rock-forming minerals. *American Mineralogist*, 68, 277–279.

Le Fort, P. (1975). Himalayas. the collided range. Present knowledge of the continental arc. *American Journal of Science*, 275A, 1–44.

Le Fort, P. and France-Lanord, C. (1995). Granites form Mustang and surrounding regions (Central Nepal). *Journal of Nepal Geological Society*, 11, 53–57.

Montemagni, C., Carosi, R., Fusi, N., Iaccarino, S., Montomoli, C., Villa, I.M., Zanchetta, S. (2020). Three-dimensional vorticity and time-constrained evolution of the Main Central Thrust zone, Garhwal Himalaya (NW India). *Terra Nova*, 32, 215–224.

Montomoli, C., Iaccarino, S., Carosi, R., Langone, A., Visonà, D. (2013). Tectonometamorphic discontinuities within the Greater Himalayan Sequence in Western Nepal (Central Himalaya). Insights on the exhumation of crystalline rocks. *Tectonophysics*, 608, 1349–1370.

Montomoli, C., Carosi, R., Iaccarino, S. (2015). Tectonometamorphic discontinuities in the Greater Himalayan Sequence: A local or a regional feature? *Geological Society of London Special Publication*, 412, 21–41.

Patel, R.C., Singh, S., Asokan, A., Manickavasagam, R.M., Jain, A.K. (1993). Extensional tectonics in the Himalayan orogen, Zanskar, NW India. In *Himalayan Tectonics*, Treloar, P.J. and Searle, M.P. (eds). Geological Society of London, Special Publications, London.

Pognante, U., Castelli, D., Benna, P., Genovese, G., Oberli, F., Meier, M., Tonarini, S. (1990). The crystalline units of the High Himalayas in the Lahul–Zanskar region (northwest India). Metamorphic–tectonic history and geochronology of the collided and imbricated Indian plate. *Geological Magazine*, 127, 101–116.

Robyr, M. and Lanari, P. (2020). Kinematic, metamorphic, and age constraints on the Miyar Thrust Zone. Implications for the Eohimalayan History of the High Himalayan Crystalline of NW India. *Tectonics*, 39, e2020TC006379.

Robyr, M., Vannay, J.C., Epard, J.L., Steck, A. (2002). Thrusting, extension, and doming during the polyphase tectonometamorphic evolution of the High Himalayan Crystalline Zone in NW India. *Journal of Asian Earth Sciences*, 21, 221–239.

Robyr, M., Hacker, B.R., Mattinson, J.M. (2006). Doming in compressional orogenic settings. New geochronological constraints from the NW Himalaya. *Tectonics*, 25.

Robyr, M., Epard, J.L., El Korh, A. (2014). Structural, metamorphic and geochronological relations between the Zanskar Shear Zone and the Miyar Shear Zone (NW Indian Himalaya). Evidence for two distinct tectonic structures and implications for the evolution of the High Himalayan Crystalline of Zanskar. *Journal of Asian Earth Sciences*, 79, 1–15.

Searle, M.P., Law, R.D., Godin, L., Larson, K.P., Streule, M.J., Cottle, J.M., Jessup, M.J. (2008). Defining the Himalayan Main Central Thrust in Nepal. *Journal of the Geological Society, London*, 165, 523–534.

Spear, F.S. (1993). *Metamorphic Phase Equilibria and Pressure-Temperature-Time Paths*. Mineralogical Society of America, Washington, DC.

Spear, F.S. (1999). Real-time AFM diagrams on your Macintosh. *Geological Materials Research*, 1(3), 1–18.

Steck, A. (2003). Geology of the NW Indian Himalaya. *Eclogae Geologicae Helvetiae*, 96, 147-U113.

Steck, A., Spring, L., Vannay, J.-C., Masson, H., Bucher, H., Stutz, E., Marchant, R., Tieche, J.-C. (1993). The tectonic evolution of the Northwestern Himalaya in eastern Ladakh and Lahul, India. *Geological Society, London, Special Publications*, 74, 265–276.

Steck, A., Epard, J.L., Vannay, J.C., Hunziker, J., Girard, M., Morard, A., Robyr, M. (1998). Geological transect across the Tso Morari and Spiti areas. The nappe structures of the Tethys Himalaya. *Eclogae Geologicae Helvetiae*, 91, 103–122.

Steck, A., Epard, J.L., Robyr, M. (1999). The NE-directed Shikar Beh Nappe. A major structure of the Higher Himalaya. *Eclogae Geologicae Helvetiae*, 92, 239–250.

Vance, D. and Harris, N. (1999). Timing of prograde metamorphism in the Zanskar Himalaya. *Geology*, 27, 395–398.

Vannay, J.C. and Grasemann, B. (2001). Himalayan inverted metamorphism and syn-convergence extension as a consequence of a general shear extrusion. *Geological Magazine*, 138, 253–276.

Vannay, J.C. and Hodges, K.V. (1996). Tectonometamorphic evolution of the Himalayan metamorphic core between the Annapurna and Dhaulaghiri, Central Nepal. *Journal of Metamorphic Geology*, 14, 635–656.

Vannay, J.C. and Steck, A. (1995). Tectonic evolution of the High Himalaya in Upper Lahul (NW Himalaya, India). *Tectonics*, 14, 253–263.

Visonà, D., Carosi, R., Montomoli, C., Peruzzo, L., Tiepolo, M. (2012). Miocene andalusite leucogranite in central-east Himalaya (Everest–Masang Kang area). Low-pressure melting during heating. *Lithos*, 144, 194–208.

Waters, D.J. (2019). Metamorphic constraints on the tectonic evolution of the High Himalaya in Nepal. The art of the possible. *Geological Society, London, Special Publications*, 483, 325–375.

6
Oligo-Miocene Exhumation of the Metamorphic Core Zone of the Himalaya Across the Range

Rodolfo CAROSI[1], Salvatore IACCARINO[1],
Chiara MONTOMOLI[1] and Martin ROBYR[2]

[1] *University of Turin, Italy*
[2] *University of Lausanne, Switzerland*

6.1. Introduction

The Himalayan orogen shows a thick metamorphic core (5–30 km) extending for nearly 2,400 km along strike (Figure 6.1). The mechanism allowing its exhumation challenged the researchers since the beginning of the study of the Himalaya. The Himalaya shows first-order faults/shear zones where deformation localized for several million years such as the South Tibetan Detachment (STDS) and Main Central Thrust zone (MCT), bounding the metamorphic core of the belt (mainly made by the Greater Himalayan Sequence: GHS) to the north and to the south, respectively, and stretching parallel to the belt for more than 2,400 km (see Volume 1 – Chapter 3).

The attention of the researchers in the last decades mainly focused on these two regional-scale faults/shear zones, which helped us to explain the Miocene exhumation of the metamorphic core giving rise to tectonic models that can be divided into two main groups: (a) models requiring the contemporaneous activity of the MCT and STDS, including models derived by numerical modeling, and (b) models not requiring the contemporaneous activity of MCT and STDS. In the light of this dualism, the factor "time" became essential to better understand and to properly decipher the timing of shearing along STDS and MCT exhumation histories of the metamorphic core.

Moreover, a multidisciplinary approach, including field work, structural analysis, metamorphic petrology and geochronology, is of primary importance, allowing us to correctly decipher deformation and structures of the GHS along the belt and their role in the exhumation history. This long, field-based work has been successful in identifying a new prominent, contractional shear zone (named High Himalayan Discontinuity: HHD; Montomoli et al. 2013, 2015) documented in several sections of the Himalaya (Figure 6.2). It was active for 11–10 Ma before the activation of the MCT and divides the GHS into two different portions: the upper portion (GHSU) is made mainly of sillimanite or kyanite-bearing migmatite, or partially molten rocks, with a higher degree of melting than the lower portion (GHSL) (Figure 6.3 and Volume 2 – Chapter 5).

The occurrence of this regional-scale tectono-metamorphic discontinuity sheds new light on the exhumation models of the GHS. However, GHS exhumation was not cylindrical and GHS in NW India behaved in a different way during exhumation by the occurrence of early NE directed tectonics and large-scale doming.

6.2. Central Himalaya

Plate Tectonics well explains, since the 1960s, how lithosphere is subducted, how the huge mass of rocks are transported down into the mantle and how the orogenic belt forms (see Volume 1 – Chapter 1). Discussion still exists to understand how deep-seated rocks return to the surface of the Earth. When driving forces cease isostatic compensation (see Volume 1 – Chapter 5), erosion (see Volume 3 – Chapter 2) and extensional tectonics are useful mechanisms to exhume deep-seated rocks in collisional belts. So that erosion and extensional tectonics allow the destruction of an orogen by removing a large amount of rocks from the upper part, and consequently deeper portions are allowed to move toward the surface.

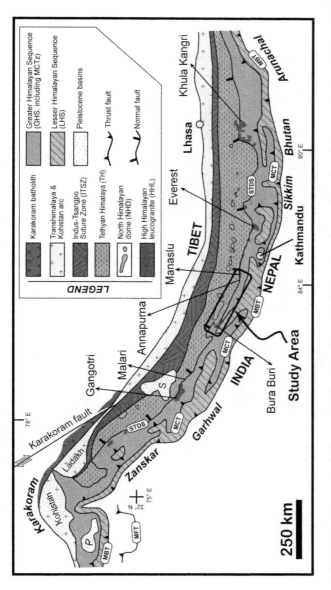

Figure 6.1. Schematic geological map of the Himalayan belt showing the main units and tectonic boundaries. MFT, Main Frontal Thrust; MBT, Main Boundary Thrust; MCT, Main Central Thrust; STDS, South Tibetan Detachment System; P, Peshawar Basin; S, Sutlej Basin (from Carosi et al. 2019, see Volume 1 – Chapter 3). For a color version of this figure, see www.iste.co.uk/cattin/himalaya2.zip

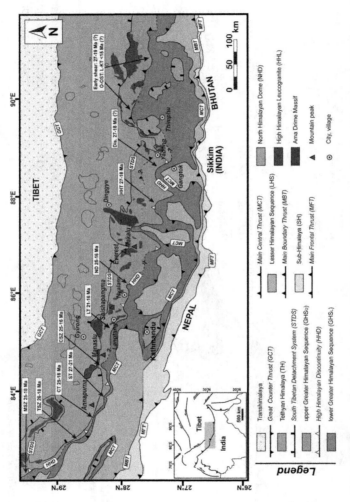

Figure 6.2. Geological sketch map of the Central and Eastern Himalaya showing the trace of the HHD from Western Nepal, through Sikkim, to Bhutan and its age (modified after Carosi et al. 2019). The GHS is divided into an upper GHS and a lower GHS by the HHD. The age of the HHD in different sections of the belt is reported in white rectangles. The location and age of the HHD are from Carosi et al. (2018, 2019) with references. For a color version of this figure, see www.iste.co.uk/cattin/himalaya2.zip

In the Himalaya, however, deep-seated rocks are exhumed when convergence in the orogen is still active (i.e. contractional forces continue to operate) and this feature cannot be easily explained with the previously described factors.

In the 1980s, the discovery of normal faults active contemporaneously, and on the same vertical, with thrusts in the Himalaya, led to propose several models to explain the exhumation of the metamorphic rocks (which are mostly included in the GHS). After this discovery, several models, requiring the contemporaneous activity of STD and MCT, have been formulated by the researchers.

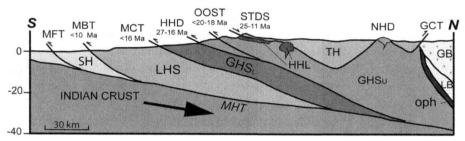

Figure 6.3. Schematic geological cross-section of the Central Himalayan belt showing the main tectonic and metamorphic discontinuities (modified after Wang et al. 2016; Carosi et al. 2018). MHT: Main Himalayan Thrust; MFT: Main Frontal Thrust; MBT: Main Boundary Thrust; MCT: Main Central Thrust; HHD: High Himalayan Discontinuity; OOST: out of sequence thrust; STD: South Tibetan Detachment; HHL: High Himalayan Leucogranite; NHD: North Himalaya Dome; GCT: Great Counter Thrust: Oph: ophiolites; LB: Lhasa Block; GB: Gandgese Batholith; SG: Siwalik Group; LHS: Lesser Himalayan Sequence; GHSL: lower Greater Himalayan Sequence; GHSU: upper Greater Himalayan Sequence: TH: Tethyan Himalaya. For a color version of this figure, see www.iste.co.uk/cattin/himalaya2.zip

The STDS, a system of normal faults and shear zones, bounds the upper part of the GHS, whereas the MCT bounds the lower part (Figures 6.2 and 6.4). The kinematics of these two faults/shear zones indicate that GHS is pushed southward for the combination of the kinematic-related arrows of the MCT and STDS (Figure 6.4(a) and (b)). At the same time, the Tethyan Himalaya, above the GHS, was translated to the north for dozen or even hundreds of kilometers due to the sliding along the STD. GHS has been considered a uniform wedge pushed up to the South, limited and guided by the two bounding faults (the lower MCT and the upper STD) for several Ma (25–23 and 16–15 Ma). This is the mechanism of wedge extrusion where GHS was initially considered as a rigid body, affected by heterogeneous simple shear, sliding along the two main bounding faults (STDS and MCT) (Figure 6.4(b)).

Subsequently, GHS has been regarded as a deformable, ductile mega-tectonic unit due to the presence of deformed metamorphic rocks. Deformation inside GHS cannot be described only in terms of simple shear, but deformation is non-coaxial, where the contemporaneous presence of simple shear (linked to the kinematic of the MCT and STDS) and pure shear (maybe due to the lithostatic load above the GHS with an additional vertical contraction) (Figure 6.4b). For this reason, extrusion is in fact a ductile extrusion where GHS underwent deformation in the ductile realm.

A variation of the extrusion model is the "wedge insertion model" proposed by Webb et al. (2007) (Figure 6.4(c)). The main difference with respect to the previous models is that the STDS is no longer regarded as a normal fault but it is regarded as a passive "backthrust" connected to the Great Counter Thrust, close to the Indus suture zone (Figure 6.1). STDS connects to the South with the MCT giving a tapered to south geometry to the GHS. In this way, GHS is a sub-horizontal wedge with a toe moving to the South.

The GHS hosts abundant migmatites and granites, especially in the upper portion. Geophysical results by Nelson et al. (1996) revealed the presence of large, melted areas below Tibet. According to Rosenberg and Handy (2005), the presence of only ca. 7–9% of melt in the rocks is enough to lower the viscosity and allow the GHS to flow. This led Beaumont et al. (2001) to formulate a numerical model at the scale of the orogen, explaining the exhumation of melted rocks by a "channel flow" where the flux of melted rocks is mainly horizontal and bounded, or guided, by the contemporaneous MCT and STD (Figure 6.4(a)). The low viscosity channel moves towards the surface in the frontal part of the orogen, leading to the exhumation of the molten rocks of the GHS due to the strong and concentrated effect of erosion by monsoon active for several Ma in the frontal part of the belt. In this way, localized, intense, erosion plays a primary role in the exhumation of metamorphic rocks. In the channel flow model deformation inside the GHS is a combination of the Poiseuille flow (also known as pipe-flow), induced by the different vertical pressures in Tibet and in the frontal part of the belt, and the Couette flow, the simple shear caused by the subduction of the Indian plate. The result of these two components is a hybrid velocity field causing the upper part of the channel to flow upward and to the South and the lower part moving northward below Tibet.

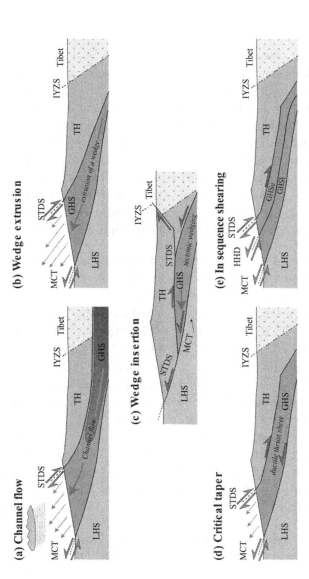

Figure 6.4. Schematic drawings of the main current tectonic models for the exhumation of GHS rocks between the Main Central Thrust (MCT) and the South Tibetan Detachment System (STDS). (a) Channel flow red: melted part of the GHS above 750°C; (b) wedge extrusion; red arrows indicate different velocities of extrusion; (c) wedge insertion; STDS is a passive back-thrust; (d) critical taper (e) in sequence shearing. IYSZ: Indus–Yarlung Suture Zone; TH: Tethyan Himalaya; GHS: Greater Himalayan Sequence; GHSU: upper Greater Himalayan Sequence; GHSL: lower Greater Himalayan Sequence; HHD: High Himalayan Discontinuity; LHS: Lesser Himalayan sequence). Models from a–c require the contemporaneous activity of MCT and STDS, whereas models from d and e are independent of the contemporaneous activity of STDS and MCT. Modified after Montomoli et al. (2013) with references therein. For a color version of this figure, see www.iste.co.uk/cattin/himalaya2.zip

The numerical models were also able to reproduce the natural P–T–t conditions of the metamorphic rock studied in Central Nepal. The GHS needs a thickness of 10–20 km and a timing of incubation of c. 10–20 Ma to start to melt and to move.

In the journey of melted rocks from below the Tibetan plateau to the surface, the temperature decreases at the point that the frontal part of the channel is cooled down below the 800–750°C isotherm and so rocks behave plastically, but are no more molten. This is called the mechanism of channel flow + ductile extrusion or channel flow + wedge insertion model Ambrose et al. (2015, and references therein). In this view, the rocks now exposed in the frontal part of the Himalaya were molten nearly ca. 20 Ma and proceed to the frontal part of the belt. GHS has a wedge shape in the frontal part. The STDS is no longer regarded as a true extensional shear zone/fault, but as a passive shear zone between GHS and TH caused by the sliding, propagation of the GHS southward. In this model, deformation is concentrated into a hot channel.

All the above-mentioned models need the contemporaneous activity of the STDS and MCT to properly work, as proposed by Godin et al. (2006). More recent field-oriented studies questioned the contemporaneous activity of STDS and MCT all along the belt in the time span of 25–16 Ma. Some strands of the STDS are older than 24–25 (Carosi et al. 2013; Iaccarino et al. 2017), whereas others are active later, even after 20 Ma. A review of the timing of the activity of the STDS along the Himalaya by Iaccarino et al. (2017) suggests a diachronicity of both STDS and MCT along the Himalayan strike, suggesting that activity on the STDS occurred in pulses not independently of the shearing along MCT. Moreover, deformation in the GHS occurred in the thrust-sense shear zone dated by U–Th–Pb on monazite at 41–36 thin portions in the uppermost part of the GHS in Central Nepal (Carosi et al. 2016), well below the requested thickness to activate the channel flow. Such old shear zones in the upper portion of the GHS leave not enough time and space to activate a channel at the whole scale of the GHS.

Only a few exhumation models for the exhumation of the GHS are independent of the contemporaneous activity of MCT and STDS. The first one proposed by Kohn (2008) is the "critical taper" model (Figure 6.4(d)). GHS has the shape of a wedge at the scale of the orogen including both the brittle and ductile portions of the belt. Kohn (2008) found that P–T–t data from Central Nepal and observations differ considerably from channel flow model's predictions both for GHS and LHS. The critical taper model includes extension at the top of the orogenic wedge (e.g. STDS) when it thickens or out of sequence thrusts, backthrusts of thrust to the foreland when the wedge is thinned. It is worth to note that the extension along the STDS is not continuous, but it is episodic in response to the overall mechanics of the orogenic wedge. In sequence thrusting and syn-deformational metamorphism are included in this model.

In the last 12 years, detailed studies in several natural sections of the Himalaya, from the LHS through the entire GHS, revealed the presence of a prominent top-to-the-South thrust-sense shear zones affecting thick portions of the metamorphic core between 28 and 17 Ma (see Volume 2 – Chapter 5). They are part of a regional-scale structure called High Himalayan Discontinuity (HHD; Montomoli et al. 2013, 2015; Carosi et al. 2018, Figures 6.2 and 6.4) documented in several sections of the GHS from Western, Central-Eastern Nepal, NW India, Sikkim and Bhutan (Figure 6.2). It was active for nearly 10 Ma and divides the GHS into GHS_U and GHS_L, the upper and lower portions of the GHS, respectively.

GHS_U is made mainly of sillimanite or kyanite-bearing migmatite, or partially molten rocks, with a higher degree of melting than the GHS_L. Due to the long-lasting activity of the HHD, different P–T paths are recorded in the GHS_U and GHS_L. Frequently, GHS_U rocks exhibit lower equilibration pressure values at Tmax, commonly in the sillimanite stability field, than the GHS_L rocks with nearly 0.2–0.3 GPa of difference. The finding of the HHD active before the MCT and in a different structural position joined to the occurrence of even older shear zones (Late Oligocene) in the upper portion of the GHS in Central Nepal allowed us to propose an in-sequence-shearing model for the exhumation of the metamorphic core (Figures 6.4a and 6.5). Upper top-to-the-South shear zones were active before shear zones in the mid-GHS (HHD). Later, deformation shifted down to the MCT (Figure 6.5). Contractional shear zones in the metamorphic core were joined to crustal thinning during syn-metamorphic deformation, erosion or may be episodic tectonic denudation (by STDS) driving the progressive exhumation of large-scale slices of the GHS.

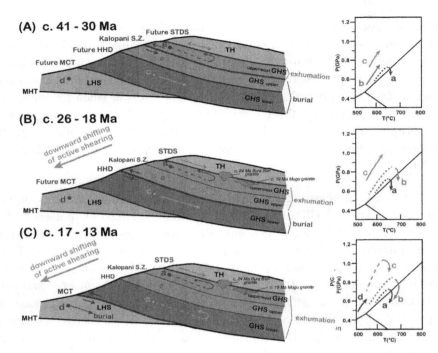

Figure 6.5. *Evolution of the Greater Himalayan Sequence (GHS) in Central Himalaya (after Carosi et al. 2018, 2019). The kinematic path of particles in the hanging wall and footwall of the shear zones is shown by dots with different colors (a–c). The graphs inserted in the right parts represent the schematic evolution of the P–T–t data at the points a–c (from Carosi et al. 2010, 2016; Montomoli et al. 2013, 2015; Iaccarino et al. 2017).* **Stage A.** *After the collisional stage, all the GHS underwent burial and consequent prograde metamorphism. At ∼41–30 Ma, the uppermost portion of the GHS (a), located above the Kalopani s.z. and below the TH, was exhuming, whereas the GHS upper and lower (b, c) was still undergoing prograde metamorphism.* **Stage B.** *At ∼26–25 Ma, following the activation of the HHD, uppermost and upper GHS (a, b), above the HHD started exhumation, whereas rocks in the footwall of the HHD (GHS lower; c,) were still buried. The older leucogranite (in orange color) intruded at ∼24–25 Ma in the upper part of the GHS cross-cutting the STDS followed by younger High Himalayan Leucogranites (Iaccarino et al. 2017, e.g. Mugu granite at 20–19 Ma).* **Stage C.** *At ∼17–13 Ma, the activation of the MCT caused the exhumation of all subunits of the GHS (a–c), and the LHS was incorporated in the belt (d), but reaching lower P and T with respect to the GHS. Only from this stage onward the GHS behaved as a unique tectonic unit. TH: Tethyan Himalaya; Kalopani s.z.: Kalopani shear zone; HHD: High Himalayan discontinuity; MCT: Main Central Thrust; LHS: Lesser Himalayan Sequence. Not to scale (modified from Carosi et al. 2018). For a color version of this figure, see www.iste.co.uk/cattin/himalaya2.zip*

Recent structural and geochronological studies in NW India revealed that the two classical portions of the MCT, the upper MCT (Vaikrita thrust) and the lower MCT (Munsiari thrust), were active at different times. U–Th–Pb on monazite and Ar/Ar on mica ages indicated the upper MCT was active from 20–16 to 9–8 Ma, whereas the lower MCT was active later at 5–4 Ma (Iaccarino et al. 2020; Montemagni et al. 2018, 2020). This fits well with an in-sequence shearing model (Carosi et al. 2016, 2018, 2019, 2022) allowing exhumation of large slices of the GHS from the upper to lower ones and involving progressively far portion of crust in the Himalayan orogeny by activation of the MCT (Figure 6.5).

6.3. North West India

Since the onset of the continental collision between Asia and India c. 55 My ago (e.g. Patriat and Achache 1984, see Volume 1 – Chapter 1), the kinematic evolution of the Himalayan range has been largely controlled by the successive extrusion of large terranes that were detached from the underthrusting Indian plate and sequentially accreted towards the south onto the orogen. In most of the Himalayan sections, the metamorphic core zone of the Himalayan range, the GHS, thrusts over the sedimentary series of the Lesser Himalaya along the MCT. This metamorphic core zone is separated from the overlying low-grade sedimentary series of the Tethyan Himalaya by the extensional structures of the STDS. Extension along the STDS and thrusting along the MCT during early Miocene suggest a tectonically controlled extrusion of the GHS towards the south. Yet, in the NW Indian Himalaya, the high-grade metamorphic rocks of the GHS are unusually exposed in an internal part of the range as a large-scale dome structure called the Gianbul dome (Figure 6.6, Dèzes et al. 1999; Robyr et al. 2014).

The Gianbul dome, which is well exposed along the Miyar and Gianbul valleys, is cored by leucogranite and migmatitic paragneiss symmetrically surrounded by rocks of the sillimanite, kyanite ± staurolite, garnet, biotite and chlorite zone. Moving towards the north, the contact between the high-grade rocks of the dome and the low-grade sediments of the Tethyan Himalaya corresponds to the Zanskar Shear Zone (ZSZ), a local equivalent of the STDS (Herren 1987; Patel et al. 1993; Dèzes et al. 1999). This NE-dipping shear zone corresponds to an Early Miocene extensional structure that initially acted as a thrust zone (Figure 6.6(b)). During Late Eocene–Early Oligocene (35-28 Ma; Vance and Harris 1999), the rocks of the GHS forming now the northern limb of the dome were underthrusted below the frontal part of the North Himalayan nappes that have affected the sedimentary series of the Tethyan Himalaya (Figure 6.7(a)). The contractional structures are superimposed by extensional structures, indicating that exhumation of the high-grade metamorphic rocks on the northern limb of the dome was controlled by extension along the ZSZ (Figure 6.7(b)). Geochronological data from various leucogranitic dykes intruding the base of the ZSZ reveal that ductile extension along the ZSZ was ongoing from at least c. 22 Ma and ceased before c. 19 Ma (Dèzes et al. 1999).

On the southern limb of the dome, along the Miyar Valley, the Miya Shear Zone (MSZ) separates the GHS high-grade rocks from the greenschist facies metasediment from the Chamba zone (Pognante et al. 1990; Steck et al. 1999; Robyr et al. 2002, 2006). It consists of a 3 km wide ductile shear zone in which shear sense criteria indicate to top-to-the NE movements. Sigmoidal inclusion trails in syntectonic garnet porphyroblasts demonstrate that the metamorphism in the Miyar valley is associated with an NE-directed crustal thickening phase.

A major feature of the tectono-metamorphic evolution of the GHS in the Miyar Valley is that the earliest phase of metamorphism and tectonism in this portion of the Himalaya relates to NE-directed thrusting. This is clearly in contrast with the southward directed thrusting and folding that has been predominant in the Himalaya since the continental collision. These unusual NE-vergent structures are associated with the emplacement of an NE-directed thrust sheet collectively termed the Shikar Beh nappe (Steck et al. 1993; Epard et al. 1995). Monazite geochronology across the MSZ yield ages ranging from 40 to 28 Ma. The structural and petrographic data indicate that the metamorphism on the southern limb of the dome results from a crustal thickening phase associated with the NE-directed propagation of crustal deformation that initiated c. 40 My ago (Figure 6.7(a)). These data indicate that the MSZ initially acted as an NE-directed synmetamorphic thrust in the frontal part of the Shikar Beh nappe along which the rocks forming now the southern limb of the dome were underthrusted below the Chamba zone.

Like the ZSZ, the MSZ was subsequently reactivated as a ductile zone of extension during Early Miocene (Figure 6.7(b)). This is supported by the occurrence of numerous top-to-the SW extensional structures such as extensional shear bands superimposed onto the contractional structures. Geochronological data of monazites from undeformed leucogranitic dykes cross-cutting the extensional structures of MSZ indicate that ductile shearing along the MSZ ended by 23.6 Ma.

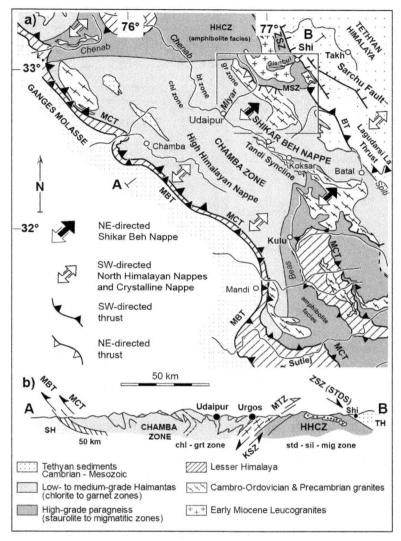

Figure 6.6. *(a) Geological map of the NW Indian Himalaya (compiled after Steck et al. 1999; Vannay and Grasemann 2001). (b) General cross-section of the NW Indian Himalaya across the Gianbul dome (modified after Steck et al. 1999). HHC = High Himalayan Crystalline, NHC = North Himalayan Crystalline, TH = Tethyan Himalaya, LH = Lesser Himalaya, SH = Sub-Himalaya, MBT = Main Boundary Thrust, MCT = Main Central Thrust, STDS = South Tibetan Detachment System, ZSZ = Zanskar Shear Zone, MSZ = Miyar Shear Zone, BT = Baralacha La Thrust*

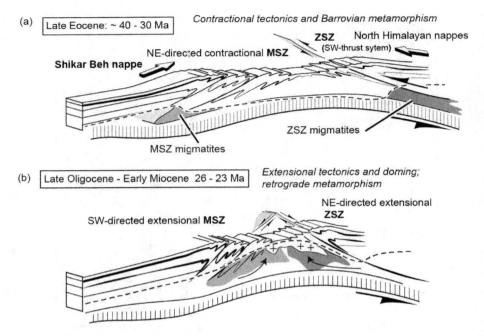

Figure 6.7. *Kinematic evolution of the GHS in Zanskar zone along the Miyar valley–Gianbul valley section. This reconstruction is based on the structural and geochronological constraints of Dèzes et al. (1999); Steck et al. (1999); Robyr et al. (2002, 2006, 2014) and Robyr and Lanari (2020)*

The geometry, metamorphic zonation and extensional tectonic contacts characterizing the Himalayan gneiss domes, like the Gianbul dome investigated in this study, are features reminiscent of metamorphic core complexes, such as those observed in the North American Cordillera, and resulting from isostatic uplift in the footwall of an extensional detachment. In contrast to classical metamorphic core complexes, however, the Himalayan gneiss domes developed within a compressional orogenic setting. The similarity between these structures does consequently not necessarily imply that they reflect a comparable doming mechanism.

Erosion is another factor that has to be taken into account in the processes of doming. Although erosion is a rather slow process of exhumation, this factor cannot be ignored as a permanent factor contributing to the exhumation processes. Moreover, in the mountainous, wet, and tectonically active regions like in the frontal part of the Himalayan range, the surficial erosion can locally be a very fast process (Avouac 2003, and Volume 3 – Chapter 2). This feature suggests a cause-and-effect relationship between the rate of erosion and the velocity of the exhumation process (e.g. Vannay et al. 2004).

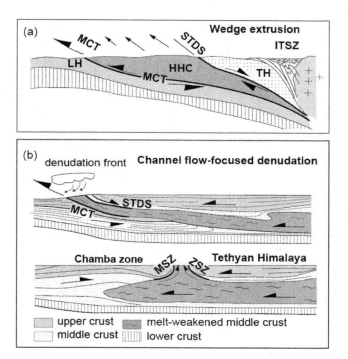

Figure 6.8. *Tectonic models for the emplacement of the GHS of Zanskar (HHCZ). (a) Kinematic model for an extrusion of the Himalayan crystalline core zone (HHC) by general shear wedge extrusion (Burchfiel and Royden 1985; Grujic et al. 1996; Vannay and Grasemann 2001; Kohn 2008). In this model, both the Main Central Thrust and South Tibetan Detachment System converge at depth. Thrusting along the MCT and coeval extension along the STDS cause a strong compression at the base of the wedge-shaped High Himalayan Crystalline generating its foreland ductile extrusion. (b) Channel flow-focused denudation model from Beaumont et al. (2001) adapted for the Himalayan range. These two models show that efficient erosion at the orogenic front and strong upper crust induce exhumation at the front of the range (upper panel), whereas reduced erosion and weaker upper crust can lead to vertical ductile extrusion in a more internal part of the range (lower panel)*

Several recent models suggest that the exhumation of the GHS, associated with combined thrusting along the MCT and extension along the STDS, reflects a ductile extrusion of these high-grade, low-viscosity paragneisses and migmatites, controlled by gravity forcing and/or by the underthrusting of the

Indian plate (e.g. Burchfiel and Royden 1982; Grujic et al. 1996; Vannay and Grasemann 2001) (Figure 6.8). These models advocate that crustal shortening is mainly accommodated by the foreland-directed extrusion of high-grade rocks as an orogenic wedge and does not account for the subvertical ductile extrusion of high-grade metamorphic rocks in a more internal part of the orogen such as that observed in the NW Himalaya (Figure 6.8).

Besides this particular feature, the GHS in the NW Himalaya is furthermore characterized by the existence of NE-vergent folds and thrust faults (MSZ) resulting from the Eohimalayan Shikar Beh nappe emplacement during Eocene. This Eocene event constitutes the key difference between the pre-exhumation history of the central and western parts of the Himalayan belt. As such, the Shikar Beh nappe emplacement may have played a crucial role in the differentiation of the kinematic evolution of the GHS during the exhumation history between the central and western parts of the Himalaya, first by modifying significantly the thermal structure within the Indian continental margin leading to the formation of a significant amount of low-density and low-viscosity migmatites in the footwall of the Shikar Beh frontal thrust and second, the NE-directed propagation of crustal deformation and the formation of the MSZ at the front of the Shikar Beh nappe considerably altered the rheological strength of the Indian continental upper crust.

According to the channel flow numerical model proposed by Beaumont et al. (2001), two significant parameters controlling the location of high-grade rock extrusion in the orogen are the erosion rate at the orogenic front and the rheological strength of the upper crustal rocks overlying the channel. These thermo-mechanical simulations suggest that efficient erosion and strong upper crust induce extrusion in the frontal part of the range, such as that observed along most of the Himalaya, whereas reduced erosion and weaker upper crust can lead to doming and extrusion in a more internal part of the range. The latter scenario appears consistent with what is observed in the NW Himalaya, where the lack of high-grade metamorphic rocks in the hanging wall of the frontal MCT, between the Beas and Chenab rivers, strongly suggests reduced erosion in the frontal part of the orogen compared to the rest of the chain, possibly because of a lack of major rivers in the Chamba area. As a result, taking advantage from the weakness of the upper crust generated by the Miyar Thrust Zone, the high-grade rocks of the GHS could have been forced to extrude in a more internal part of the orogen, to eventually "pierce" through the Tethyan Himalaya sedimentary cover as large-scale domes.

This model therefore suggests that the overburden of metasedimentary rocks lying in the foreland and weighing down on it as the extrusion of the GHS was ongoing acted as a backstop for the ductile exhumation of the GHS rocks and that decompression led to a positive feedback with partial melting and enhanced exhumation.

6.4. Conclusion

In a first-order orogen, like the Himalaya, which extends for nearly 2,400 km from west to east, the mechanism of exhumation of the GHS high-grade metamorphic rocks varies along strike. In the central part that corresponds to the more linear part of the orogeny, the exhumation is mainly controlled and facilitated by ductile shear zones arranged in an in-sequence way, so that shear zones are progressively activated southward, towards the foreland. Moving to the NW part of the belt, the orogen becomes progressively curved, approaching the Nanga Parbat syntaxis. The change in regional stresses and spatial distribution of the major compressional structures leads to an exhumation mainly accommodated by large-scale doming. In a more general way, detailed studies along different Himalayan sections show that the mechanism of exhumation is largely controlled by the arrangement of the shear zones, sometimes preexisting, developed starting from the Eohimalayan tectono-metamorphic compressional phases.

6.5. References

Ambrose, T.K., Larson, K.P., Guilmette, C., Cottle, J.M., Buckingham, H., Rai, S. (2015). Lateral extrusion, underplating, and out-of-sequence thrusting within the Himalayan metamorphic core, Kanchenjunga, Nepal. *Lithosphere*, 7, 441–464.

Avouac, J.P. (2003). Mountain building, erosion, and the seismic cycle in the Nepal Himalaya. In *Advances in Geophysics*, Dmowska, R. (ed.). Elsevier, Amsterdam.

Beaumont, C., Jamieson, R.A., Nguyen, M.H., Lee, B. (2001). Himalayan tectonics explained by extrusion of a low-viscosity crustal channel coupled to focused surface denudation. *Nature*, 414, 738–742.

Burchfiel, B.C. and Royden, L. (1982). Carpathian foreland fold and thrust belt and its relation to Pannonian and other basins. *AAPG Bulletin-American Association of Petroleum Geologists*, 66, 1179–1195.

Burchfiel, B.C. and Royden, L. (1985). North-south extension within the convergent Himalayan region. *Geology*, 13, 679–682.

Carosi, R., Montomoli, C., Rubatto, D., Visonà, D. (2010). Late Oligocene high-temperature shear zones in the core of the Higher Himalayan Crystallines (Lower Dolpo, Western Nepal). *Tectonics*, 29, TC4029.

Carosi, R., Montomoli, C., Rubatto, D., Visonà, D. (2013). Leucogranite intruding the South Tibetan Detachment in western Nepal. Implications for exhumation models in the Himalayas. *Terra Nova*, 25(6), 478–489.

Carosi, R., Montomoli, C., Iaccarino, S., Massonne, H.-J., Rubatto, D., Langone, A., Gemignani, L., Visonà, D. (2016). Middle to late Eocene exhumation of the Greater Himalayan Sequence in the Central Himalayas. Progressive accretion from the Indian plate. *Geological Society of America Bulletin*, 128(11–12), 1571–1592.

Carosi, R., Montomoli, C., Iaccarino, S. (2018). 20 years of geological mapping of the metamorphic core across Central and Eastern Himalayas. *Earth-Science Reviews*, 177, 124–138.

Carosi, R., Montomoli, C., Iaccarino, S., Visonà, D. (2019). Structural evolution, metamorphism and melting in the Greater Himalayan Sequence in central-western Nepal. *Geological Society, London, Special Publications*, 483, 305–323.

Carosi, R., Montomoli, C., Iaccarino, S., Montemagni, C., Benetti, B. (2022). A review of localization of the deformation in Garhwal Himalaya: Younging activation of shear zones from the metamorphic core of the Belt to the Foreland. *Himalayan Geology*, 43, 221–230.

Dèzes, P.J., Vannay, J.C., Steck, A., Bussy, F., Cosca, M. (1999). Synorogenic extension. Quantitative constraints on the age and displacement of the Zanskar shear zone (northwest Himalaya). *Geological Society of America Bulletin*, 111, 364–374.

Epard, J.L., Steck, A., Vannay, J.C., Hunziker, J. (1995). Tertiary Himalayan structures and metamorphism in the Kulu valley (Mandi-Khoksar transect of the western Himalaya) – Shikar Beh Nappe and Crystalline Nappe. *Schweizerische Mineralogische Und Petrographische Mitteilungen*, 75, 59–84.

Godin, L., Grujic, D., Law, R.D., Searle, M.P. (2006). Channel flow, ductile extrusion and exhumation in continental collision zones: An introduction. In *Channel Flow, Ductile Extrusion and Exhumation in Continental Collision Zones*, Law, R.D., Searle, M.P., Godin, L. (eds). Geological Society, Special Publications, London.

Grujic, D., Casey, M., Davidson, C., Hollister, L.S., Kundig, R., Pavlis, T., Schmid, S. (1996). Ductile extrusion of the Higher Himalayan Crystalline in Bhutan: Evidence from quartz microfabrics. *Tectonophysics*, 260, 21–43.

Herren, E. (1987). Zanskar shear zone – Northeast-southwest extension within the higher Himalayas (Ladakh, India). *Geology*, 15, 409–413.

Iaccarino, S., Montomoli, C., Carosi, R., Montemagni, C., Massonne, H.-J., Langone, A., Jain, A.K., Visonà, D. (2017). Pressure-temperature-deformation-time constraints on the South Tibetan detachment system in Garhwal Himalaya (NW India). *Tectonics*, 36, 2281–2304.

Iaccarino, S., Montomoli, C., Montemagni, C., Massonne, H.-J., Langone, A., Jain, A.K., Visonà, D., Carosi, R. (2020). The Main Central Thrust zone along the Alaknanda and Dhauli Ganga valleys (Garhwal Himalaya, NW India): Insights into an inverted metamorphic sequence. *Lithos*, 372, 105669.

Kohn, M.J. (2008). P-T-t data from central Nepal support critical taper and repudiate large-scale channel flow of the Greater Himalayan Sequence. *Geological Society of America Bulletin*, 120, 259–273.

Kündig, R. (1989). Domal structures and high-grade metamorphism in the Higher Himalayan Crystalline, Zanskar region, Northwest Himalaya, India. *Journal of Metamorphic Geology*, 7, 43–55.

Montemagni, C., Iaccarino, S., Montomoli, C., Carosi, R., Jain, A.K., Villa, I.M. (2018). Age constraints on the deformation style of the South Tibetan Detachment System in Garhwal Himalaya. *Italian Journal of Geosciences*, 137, 175–187.

Montemagni, C., Carosi, R., Fusi, N., Iaccarino, S., Montomoli, C., Villa, I.M., Zanchetta, S. (2020). Three-dimensional vorticity and time-constrained evolution of the Main Central Thrust zone, Garhwal Himalaya (NW India). *Terra Nova*, 32, 215–224.

Montomoli, C., Iaccarino, S., Carosi, R., Langone, A., Visonà, D. (2013). Tectonometamorphic discontinuities within the Greater Himalayan Sequence in Western Nepal (Central Himalaya): Insights on the exhumation of crystalline rocks. *Tectonophysics*, 608, 1349–1370.

Montomoli, C., Carosi, R., Iaccarino, S. (2015). Tectonometamorphic discontinuities in the Greater Himalayan Sequence: A local or a regional feature? In *Tectonics of the Himalaya*, Mukherjee S., van der Beek, P., Mukherjee, P.K. (eds). Geological Society, Special Publication, London.

Nelson, K.D., Zhao, W., Brown, L.D., Kuo, J., Che, J., Liu, X., Klemperer, S.L., Makovsky, Y., Meissner, R.J.J.M., Mechie, J. et al. (1996). Partially molten middle crust beneath southern Tibet: Synthesis of project INDEPTH results. *Science*, 274(5293), 1684–1688.

Patel, R.C., Singh, S., Asokan, A., Manickavasagam, R.M., Jain, A.K. (1993). Extensional tectonics in the Himalayan orogen, Zanskar, NW India. In *Himalayan Tectonics*, Treloar, P.J. and Searle, M.P. (eds). Geological Society, Special Publications, London.

Patriat, P. and Achache, J. (1984). India–Eurasia collision chronology has implications for crustal shortening and driving mechanisms of plates. *Nature*, 311, 615–621.

Pognante, U., Castelli, D., Benna, P., Genovese, G., Oberli, F., Meier, M., Tonarini, S. (1990). The crystalline units of the High Himalayas in the Lahul–Zanskar region (northwest India): Metamorphic–tectonic history and geochronology of the collided and imbricated Indian plate. *Geological Magazine*, 127, 101–116.

Robyr, M. and Lanari, P. (2020). Kinematic, metamorphic, and age constraints on the Miyar Thrust Zone: Implications for the Eohimalayan History of the High Himalayan Crystalline of NW India. *Tectonics*, 39, e2020TC006379.

Robyr, M., Vannay, J.C., Epard, J.L., Steck, A. (2002). Thrusting, extension, and doming during the polyphase tectonometamorphic evolution of the High Himalayan Crystalline Zone in NW India. *Journal of Asian Earth Sciences*, 21, 221–239.

Robyr, M., Hacker, B.R., Mattinson, J.M. (2006). Doming in compressional orogenic settings: New geochronological constraints from the NW Himalaya. *Tectonics*, 25,

Robyr, M., Epard, J.L., El Korh, A. (2014). Structural, metamorphic and geochronological relations between the Zanskar Shear Zone and the Miyar Shear Zone (NW Indian Himalaya): Evidence for two distinct tectonic structures and implications for the evolution of the High Himalayan Crystalline of Zanskar. *Journal of Asian Earth Sciences*, 79, 1–15.

Rosenberg, C.L. and Handy M.R. (2005). Experimental deformation of partially melted granite revisited: Implications for the continental crust. *Journal of Metamorphic Geology*, 23, 19–28.

Steck, A., Spring, L., Vannay, J.-C., Masson, H., Bucher, H., Stutz, E., Marchant, R., Tieche, J.-C. (1993). The tectonic evolution of the Northwestern Himalaya in eastern Ladakh and Lahul, India. *Geological Society, London, Special Publications*, 74, 265–276.

Steck, A., Epard, J.L., Robyr, M. (1999). The NE-directed Shikar Beh Nappe: A major structure of the Higher Himalaya. *Eclogae Geologicae Helvetiae*, 92, 239–250.

Vance, D. and Harris, N. (1999). Timing of prograde metamorphism in the Zanskar Himalaya. *Geology*, 27, 395–398.

Vannay, J.C. and Grasemann, B. (2001). Himalayan inverted metamorphism and syn-convergence extension as a consequence of a general shear extrusion. *Geological Magazine*, 138, 253–276.

Vannay, J.C., Grasemann, B., Rahn, M., Frank, W., Carter, A., Baudraz, V., Cosca, M. (2004). Miocene to Holocene exhumation of metamorphic crustal wedges in the NW Himalaya: Evidence for tectonic extrusion coupled to fluvial erosion. *Tectonics*, 23(1).

Wang, J.M., Zhang, J.J., Liu, K., Zhang, B., Wang, X.X., Rai, S.M., Scheltens, M. (2016). Spatial and temporal evolution of tectonometamorphic discontinuities in the central Himalaya: Constraints from P–T paths and geochronology. *Tectonophysics*, 679, 41–60.

Webb, A.A.G., Yin, A., Harrison, T.M., Célérier, J., Burgess, W.P. (2007). The leading edge of the Greater Himalayan crystallines revealed in the NW Indian Himalaya: Implications for the evolution of the Himalayan Orogen. *Geology*, 35, 955–958.

PART 3

Lesser and Sub Himalayan Sequence

7

Lithostratigraphy, Petrography and Metamorphism of the Lesser Himalayan Sequence

Chiara GROPPO[1,2], Franco ROLFO[1,2],
Shashi TAMANG[3,4,5] and Pietro MOSCA[2]
[1]*Department of Earth Sciences, University of Turin, Italy*
[2]*Institute of Geoscience and Geohazard, Turin, Italy*
[3]*University of Turin, Italy*
[4]*Paris Cité University, France*
[5]*Tribhuvan University, Kathmandu, Nepal*

7.1. Introduction

The Lesser Himalayan Sequence (LHS) is a thick (more than 15–20 km), mostly Paleo- to Meso-Proterozoic sedimentary sequence originally deposited on the northern margin of the Indian plate (e.g. Gansser 1964; Parrish and Hodges 1996; DeCelles et al. 2000; Martin et al. 2005), intruded by Proterozoic igneous lithologies (e.g. Kohn et al. 2010; Larson et al. 2017, 2019). During the Himalayan orogeny, this Proterozoic sedimentary sequence was variably deformed and metamorphosed developing a typical inverted metamorphism, with metamorphic grade increasing from south to north and from lower to upper structural levels (e.g. Pêcher 1989; Kohn 2014).

In Nepal, the LHS is mostly exposed in the midlands, at altitudes ranging from 300 to 3,500 m a.s.l., and forms a belt up to 100 km wide (Figure 7.1). The current LHS exposure is due to the interplay between thrusting, folding and erosion. The structure of the LHS is dominated by a regional-scale structural culmination (i.e. the Lesser Himalayan Duplex: DeCelles et al. 2001; Pearson and DeCelles 2005), testified by the occurrence of a broad antiform developed longitudinally with respect to the Himalayan thrust-belt. The erosion was more intense towards the west (Dithal 2015), resulting in a wider exposure of the LHS in western and central Nepal compared to eastern Nepal; there, the LHS crops out in large antiformal tectonic windows (in the Chautara-Okhaldhunga, Arun valley and Tamur valley regions) (Figure 7.1).

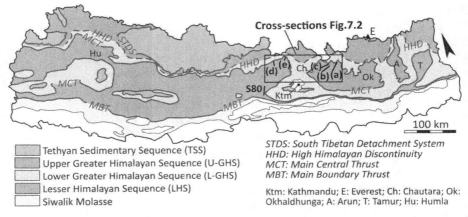

Figure 7.1. *Geological sketch map of the Nepal Himalaya, showing the major tectono-metamorphic units (modified from Dithal 2015; He et al. 2015). The location of the High Himalayan Discontinuity (HHD) is based on: Groppo et al. (2009, 2010, 2012, 2013), Mosca et al. (2012) and Rapa et al. (2016, 2018) for central-eastern Nepal and on Montomoli et al. (2013) and Wang et al. (2016) for central-western Nepal. The inset locates the cross sections reported in Figure 7.2; S80: reference section of Stöcklin (1980). For a color version of this figure, see www.iste.co.uk/cattin/himalaya2.zip*

In the Himalayan fold-thrust belt, the LHS is tectonically interposed between the underlying Siwalik Group (molasses) to the south (see Volume 2 – Chapter 8) and the overlying Greater Himalayan Sequence (GHS) to the north (see Volume 2 – Chapters 5 and 6), from which it is separated by the Main Boundary Thrust (MBT) and the Main Central Thrust (MCT), respectively (Yin and Harrison 2000) (Figure 7.1). The precise location of the main tectonic discontinuity (i.e. the MCT) juxtaposing the GHS over the LHS is still debated, because a number of different criteria (structural, metamorphic, chronological, compositional; see, for example, Searle et al. (2008) for a review) have been used to define it. Here, we follow the definition proposed by Goscombe et al. (2006, 2018), namely: (i) the boundary between the LHS and GHS is mostly a lithostratigraphic boundary (i.e. Himalayan Unconformity HU), separating two sequences with different and unique provenance; (ii) rather than being a discrete thrust, the MCT is a broad high strain zone (i.e. Main Central Thrust Zone: MCTZ) which affects both the upper structural levels of the LHS and the lower structural levels of the GHS; (iii) the lower boundary of the MCTZ is marked by two distinct tectonic discontinuities, recognized all along the Himalayan belt: the Main Central Thrust (or Basal Main Central Thrust according to Goscombe et al. (2018)) at upper structural levels, mostly coincident with the HU and the Ramgarh Thrust (as defined in central-western Nepal by DeCelles et al. (2000) and Robinson et al. (2001, 2006)) or Munsiari Thrust Valdiya (in Garhwal, NW India; 1980)) at lower structural levels.

At a regional scale, two main domains are generally distinguished within the LHS (e.g. Dithal 2015): a narrow outer (southern) domain consisting of predominantly non-metamorphosed sedimentary rocks, and a broad inner (northern) domain, recording a metamorphic overprint whose grade increases upward from the biotite to the garnet and the staurolite/kyanite first appearance (e.g. Kohn 2014). This chapter focuses on the metamorphic successions of LHS in Nepal Himalaya; therefore, the outer, non-metamorphic, LHS domain is not discussed in the following.

7.2. Lithostratigraphy and petrography

The stratigraphic reconstruction of the extremely thick (more than 15 km) LHS metasedimentary sequence along the approximately 750 km long stretch of LHS in Nepal is prevented by its complex structural setting (DeCelles et al. 2001; Pearson and DeCelles 2005). Different stratigraphic units have been defined in different areas, whose correlation at the regional scale is not always straightforward. In western and central Nepal, the stratigraphic subdivisions of the LHS are mostly based on the detailed mapping by Stöcklin (1980), Sakai (1983), Colchen et al. (1986), DeCelles et al. (2001) and Robinson et al. (2001), whereas in eastern Nepal, the LHS stratigraphy remains poorly understood and mostly based on data from Schelling and Arita (1991) and Schelling (1992). In this chapter, the lithostratigraphic and petrographic features of the LHS are described starting from the reference section defined by Stöcklin (1980) and Upreti (1999) in central Nepal (S80 in Figure 7.1) and combining our own observations from several sections in central Nepal (Figure 7.2); the main along-strike variations with respect to these reference sections are then discussed.

The LHS is conventionally divided in two complexes, characterized by distinct lithological associations: the Lower-LHS (Lower Nawakot Unit; Stöcklin 1980), dominated by siliciclastic rocks, and the Upper-LHS (Upper Nawakot Unit; Stöcklin 1980) consisting of both carbonatic and siliciclastic rocks. U-Pb dating of detrital zircons constrained the depositional age of the Lower-LHS to the Paleo-Proterozoic, with most ages converging in the interval 1,900–1,850 Ma (Parrish and Hodges 1996; DeCelles et al. 2000; Martin et al. 2005, 2011; Sakai et al. 2013), whereas the depositional age of the Upper-LHS sequence is still poorly constrained, although a Paleo- to Meso-Proterozoic age is tentatively proposed in the literature (e.g. DeCelles et al. 2000; Martin et al. 2005; Gehrels et al. 2011).

7.2.1. *Lower-LHS*

The Lower-LHS sequence consists of a thick package of metasedimentary rocks mostly derived from pelitic and arenaceous protoliths. From the bottom to the top of the sequence, the following formations have been distinguished in the reference section of Stöcklin (1980): Kuncha Formation, Fagfog Quartzite Formation and Dandagon Phyllite Formation.

7.2.1.1. *Kuncha Formation*

The Kuncha Formation is a thick (more than 2,500 m), monotonous succession of gray-greenish phyllites, meta-sandstones and minor meta-conglomerates, alternating in layers of variable thickness (from decimetric to decametric) (Figure 7.3 (a)–(c)), derived from the metamorphism of original turbiditic flysch deposits.

In the fine-grained **phyllites**, the pervasive main foliation is defined by muscovite ± chlorite ± biotite, this last mostly confined to the quartz-richer domains. The main foliation derives from the transposition of an earlier foliation, still preserved in microlithons (Figure 7.3(d)), and it is often pervasively crenulated, with the local development of an axial-plane foliation, defined by muscovite ± biotite. Small garnet grains (Figure 7.3(d)) locally appear in the uppermost (northern) part of this formation.

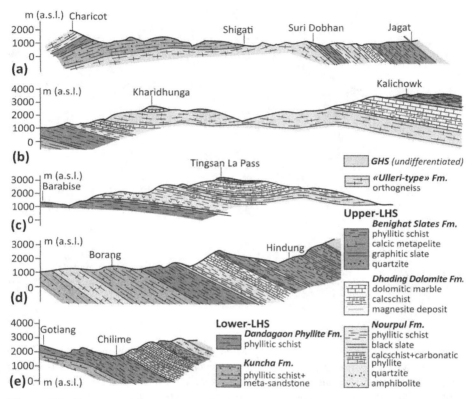

Figure 7.2. *Representative cross-sections across the Lower- and Upper-LHS in central Nepal (unpublished data from the authors). Location of each cross-section is reported in Figure 7.1. Most pictures reported in Figures 7.3–7.12 refer to these cross-sections or to adjacent transects, except where explicitly indicated. For a color version of this figure, see www.iste.co.uk/cattin/himalaya2.zip*

The **meta-sandstones** show a well-preserved clastic structure, with abundant millimeter-sized clasts of quartz and minor feldspars and muscovite (Figure 7.3(e)), already evident at the outcrop scale. The foliation is less pervasive than in the associated phyllites, resulting in a massive appearance at the outcrop scale (Figure 7.3(c)), and it is defined by the alignment of muscovite, biotite and chlorite. Chloritoid sporadically occurs in both the phyllites and the meta-sandstones; in the former, it forms millimeter-sized porphyroblasts syn- to post-kinematic with respect to the main foliation, whereas in the latters, it forms fine-grained radial aggregates.

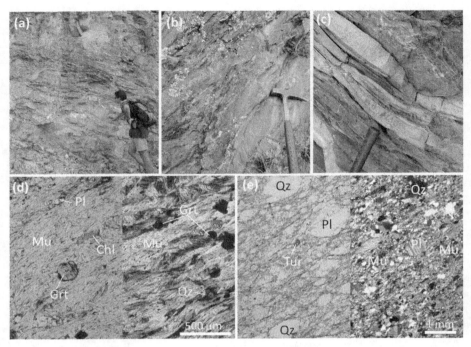

Figure 7.3. *Kuncha Formation. (a–c) Meta-sandstone intercalations (whitish and massive) within gray-greenish, pervasively foliated, phyllites. (d) Garnet-bearing muscovite + chlorite phyllite (left: Plane Polarized Light, PPL; right: Crossed Polarized Light, XPL). (e) Meta-sandstone preserving the clastic structure of the protolith (PPL/XPL). (c) and (d) are from the Tamur and Arun sections in eastern Nepal (Figure.7.1). Minerals abbreviations according to Whitney and Evans (2010). For a color version of this figure, see www.iste.co.uk/cattin/himalaya2.zip*

7.2.1.2. *Fagfog quartzite formation*

The Fagfog Quartzite Formation consists of fine- to medium-grained banded quartzite (Figure 7.4(b)) with several phyllite intercalations. This formation reaches a maximum thickness of approximately 500–800 m, but it disappears in some sections. Its stratigraphic position is controversial, having been placed either below (Stöcklin 1980) or above (Shrestha et al. 1987) the Dandagaon Phyllites Formation. According to our own observations in the inner (northern) LHS domain in central Nepal, a thick quartzite layer often occurs stratigraphically above the Dandagaon Phyllite, but its attribution to the

Fagfog Quartzite Formation is ambiguous, and it is consequently not included as a stand-alone formation in Figure 7.2.

The **quartzites** show a marked fissility and tend to separate in dm-thick plates (Figure 7.4(a)), although the foliation, defined by muscovite and minor biotite, is not pervasive (Figure 7.4(c)).

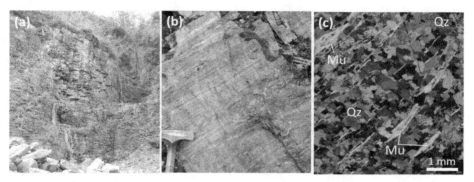

Figure 7.4. *Fagfog Quartzite Formation. (a) and (b) Outcrop appearance of banded, highly fissile quartzite. (c) The main foliation in the quartzite is defined by muscovite (XPL). For a color version of this figure, see www.iste.co.uk/cattin2/himalaya2.zip*

7.2.1.3. Dandagaon Phyllite Formation

The Dandagaon Phyllite Formation is described as a succession dominated by dark gray and gray-green homogeneous phyllites, with sporadic thin layers of quartzites. Compared to the Kuncha Formation, the Dandagaon Phyllite Formation is almost free of meta-sandstone intercalations. Centimeter-thick quartz veins, deformed and stretched parallel to the main foliation, are locally quite abundant at this stratigraphic level (Figure 7.5(a) and (b)). An additional common feature in this formation is the presence of a pronounced mineral lineation defined by biotite (Figure 7.5(c)).

In the **phyllites**, the main foliation, defined by muscovite ± chlorite ± biotite, transposes an earlier foliation preserved in the microlithons (Figure 7.5(d) and (e)) and it is often pervasively crenulated. Pre-kinematic garnet porphyroblasts (up to a few millimeters in size) wrapped around by the main foliation are common (Figure 7.5(d)), as well as late biotite porphyroblasts overgrowing the main foliation (Figure 7.5(f)), which are not related to the mineral lineation described above.

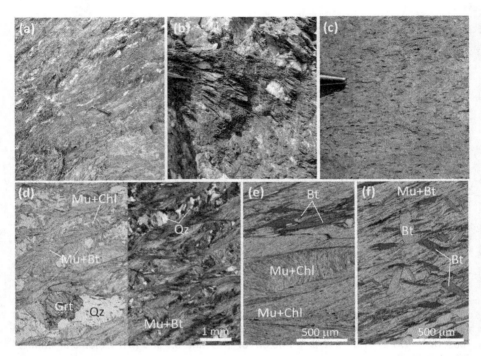

Figure 7.5. Dandagaon Phyllite Formation. (a) and (b) Quartz veins stretched parallel to the main foliation. (c) Mineral lineation defined by biotite in a garnet-bearing phyllite. (d) Pre-kinematic garnet porphyroblast wrapped around by a foliation defined by muscovite + chlorite + biotite (PPL/XPL). (e) The main foliation derives from transposition of an early foliation preserved in the microlithons (PPL). (f) Late biotite flakes statically overgrowing the main foliation (PPL). For a color version of this figure, see www.iste.co.uk/cattin/himalaya2.zip

7.2.2. *Upper-LHS*

The Upper-LHS sequence differs from the Lower-LHS for its marked lithological heterogeneity and for the occurrence of carbonate-bearing lithologies, whose abundance is, nevertheless, highly variable along strike. In the central Nepal reference section (Stöcklin 1980), the following formations have been distinguished from the bottom to the top of the sequence: Nourpul, Dhading Dolomite, Benighat Slate, Malekhu Limestone and Robang Phyllite Formations.

Lithostratigraphy, Petrography and Metamorphism   167

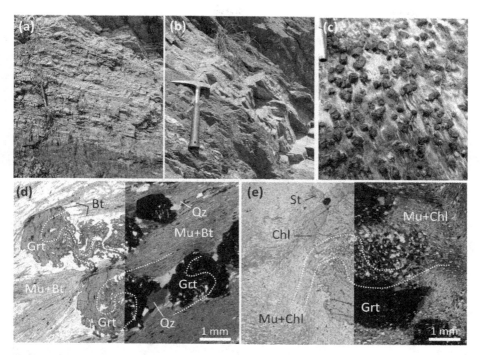

Figure 7.6. *Nourpul Formation. (a) and (b) Banded appearance of the Nourpul Formation, with alternating layers consisting of different lithologies. From bottom to top: (a) graphitic phyllites, calcschists and impure marbles; (b) graphitic phyllites, quartzites and phyllites. (c) Millimeter-sized garnet porphyroblasts in a phyllite. (d) and (e) Syn-kinematic garnet porphyroblasts including a rotated internal foliation continuous with the external one, defined by muscovite + biotite (d) or by muscovite + chlorite ± biotite (e). Garnet rims overgrow the main foliation (PPL/XPL). For a color version of this figure, see www.iste.co.uk/cattin/himalaya2.zip*

7.2.2.1. *Nourpul Formation*

The Nourpul Formation is a lithologically heterogeneous succession mostly consisting of phyllites, black slates, carbonatic phyllites, calcschists and minor quartzites, alternated in metric to decametric layers. At the outcrop scale this formation typically appears thickly banded and varied in color (Figure 7.6(a) and (b)); the heterogeneity at the outcrop scale is the main feature which allows distinguishing this formation from the underlying Dandagaon Phyllite Formation. Protoliths of these lithologies are mostly represented by pelites variably enriched in carbonaceous material, dolomitic pelites, dolomitic marls and quartzitic sandstones. Quartzites are generally concentrated in the lowermost portion of the formation, whereas the amount of carbonatic lithologies increases upward. Thin layers and/or small lenticular bodies (up to a few meters in thickness) of amphibolites, possibly derived from original basic dykes (Figure 7.7(c)), occasionally occur in this formation.

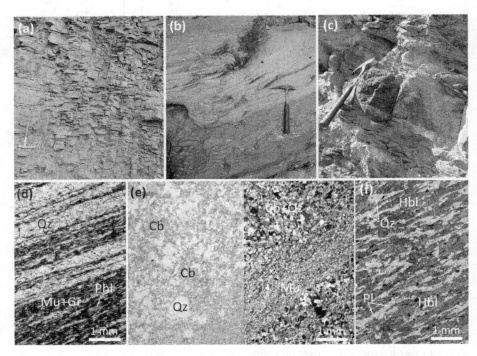

Figure 7.7. Nourpul Formation. (a) Black slates. (b) Banded calcschists. (c) Thin layer of amphibolites within phyllites. (d) Black slate: the foliation is defined by muscovite, graphite and phlogopite (PPL). (e) Calcschist: the foliation is defined by muscovite, concentrated in millimeter-thick layers (PPL/XPL). (f) Amphibolite: the foliation is defined by hornblende (PPL). For a color version of this figure, see www.iste.co.uk/cattin/himalaya2.zip

In the **phyllites**, the main foliation is generally defined by abundant muscovite and minor biotite, more rarely also by chlorite. They typically contain syn- to post-kinematic garnet porphyroblasts, up to several millimeters in size (Figure 7.6(c)–(e)); staurolite porphyroblasts are more rare and they generally overgrow the main foliation (Figure 7.6(e)).

Black slates (Figure 7.7(a)) have a very simple assemblage consisting of abundant graphite, quartz, phlogopite and/or muscovite, the last two aligned to define the main foliation (Figure 7.7(d)).

Calcschists and impure marbles are intimately associated and generally show a banded structure, with alternating layers characterized by different proportions of the silicatic and carbonatic fractions (Figure 7.7(b)); the main foliation is defined by phlogopite and/or muscovite (Figure 7.7(e)).

Amphibolites are generally characterized by a relatively simple mineral assemblage, consisting of peciloblastic hornblende, quartz and plagioclase (Figure 7.7(f)), and accessory ilmenite and/or titanite. Epidote and garnet only rarely occur within this lithology.

7.2.2.2. Dhading Dolomite Formation

The transition from the Nourpul Formation to the overlying Dhading Dolomite Formation is marked by a progressive increase of the carbonatic lithologies counterbalanced by a decrease of the siliciclastic ones. However, the transition is gradual and the distinction of the two formations is not always straightforward. The Dhading Dolomite Formation mostly consists of fine-grained phlogopite ± muscovite-bearing dolomitic marbles with intercalations of calcschists, carbonatic phyllites and minor graphitic schists, respectively, derived from the metamorphic transformation of original dolostones, dolomitic marls, dolomitic pelites and pelites enriched in carbonceous material.

In spite of their metamorphic recrystallization, the light to dark gray **dolomitic marbles** generally preserve a finely laminated structure (Figure 7.8(a)), possibly inherited from the dolostone protolith. Remnants of stromatolitic structures are widely reported from several localities in central and western Nepal (Figure 7.8(c)). In the dolomitic marbles, the main foliation is defined by the preferred orientation of phlogopite and/or minor muscovite (Figure 7.8(d)). Sulfide-rich layers are locally present (Figure 7.8(b) and (e)), as well as radial aggregates of acicular tremolite developed at the expenses of coarse-grained phlogopite.

The mineral assemblage in the **calcschists** is similar to that of the marbles, but with a more abundant silicatic fraction; the main foliation is defined either by a light orange phlogopite or by a greenish biotite associated with minor muscovite (Figure 7.8(f)). **Graphitic schists** are dominated by graphite and quartz, with phlogopite defining the main foliation.

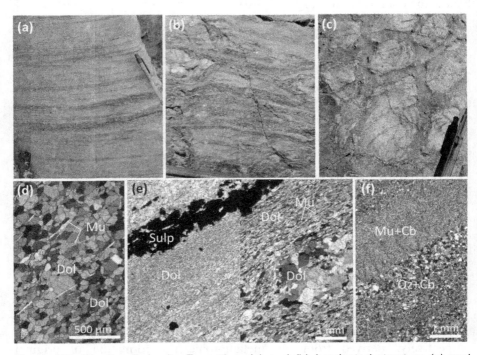

Figure 7.8. *Dhading Dolomite Formation. (a) and (b) Laminated structure (a) and sulfide-rich layers (b) in dolomitic marbles. (c) Preserved stromatolitic structure (Humla section, western Nepal; Figure 7.1). (d) Microstructure of a dolomitic marble, with a discontinuous foliation defined by muscovite (XPL). (e) Dolomitic marble with sulfide-rich layers (PPL/XPL). (f) Microstructure of a calcschist (XPL). For a color version of this figure, see www.iste.co.uk/cattin/himalaya2.zip*

Massive **magnesite + talc ores** occur as lenses and layers, from a few meters to tens of meters thick, associated with the dolomitic marbles and/or to the graphitic schists (Figure 7.9(a)). The largest magnesite + talc deposit is exposed in the Kharidhunga area of central Nepal (Figure 7.2(b)) and has been exploited for many years. Magnesite occurs as coarse-grained centimetric crystals, locally with a columnar habit (Figure 7.9(d)) or forming radial, rosette-type aggregates. Coarse-grained magnesite is partially replaced by finer-grained magnesite, talc and minor Mg-chlorite (Figure 7.9(e) and (f)). Talc veins, from a few mm to a few cm thick (Figure 7.9(c)) locally form a pervasive network (Figure 7.9(b)). The origin of these sparry magnesite deposits has been attributed to syn-sedimentary diagenetic replacement of dolomite by magnesite in an evaporitic environment developed behind biohermal barriers which inhibited water circulation (Valdiya 1995; Joshi and Sharma 2015). Talc and Mg-chlorite replaced magnesite through metamorphic hydration reactions.

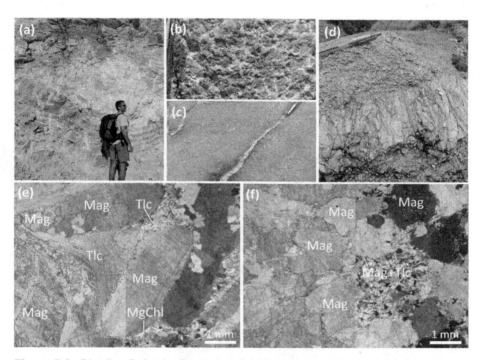

Figure 7.9. *Dhading Dolomite Formation. (a) Magnesite + talc lens (white) embedded within dolomitic marbles (yellowish). (b) and (c) Talc veins forming a pervasive inter-granular network (b) or cross-cutting the magnesite rock (c). (d) Coarse-grained magnesite with a columnar habit. (e) and (f) Microstructure of the sparry magnesite deposits, with talc and minor Mg-chlorite replacing magnesite through a pervasive network of veins (PPL/XPL). For a color version of this figure, see www.iste.co.uk/cattin/himalaya2.zip*

7.2.2.3. *Benighat Slate Formation*

The Benighat Slate Formation is the thickest formation of the Upper-LHS, ranging in thickness from 500 to more than 3,000 m. It is lithologically heterogeneous and dominated by dark phyllites, with frequent intercalations of graphitic schists (Figure 7.10(a) and (b)) and calcic metapelites, and minor carbonatic lithologies (calcschists and impure marbles) and quartzites

(Figure 7.11(a)–(c)). Protoliths of these lithologies are mostly represented by pelites enriched in carbonaceous material, dolomitic pelites and dolomitic marls, with minor dolostones and limestones.

The most common lithology is represented by two-micas, graphitic **phyllites** with porphyroblasts of aluminous minerals (garnet, staurolite and/or kyanite) (Figure 7.10(c) and (d)). Garnet porphyroblasts are typically syn- to post-kinematic, and can reach several millimeter in size (Figure 7.10(e)); they locally show large rim domains statically overgrowing the main foliation (Figure 7.10(f) and (g)), sometimes already evident at the outcrop scale (Figure 7.10(d)). Staurolite and kyanite are generally post-kinematic (Figure 7.10(f) and (g)) and can be up to a few centimeters long. Large porphyroblasts of biotite locally overgrow the main foliation (Figure 7.10(h)).

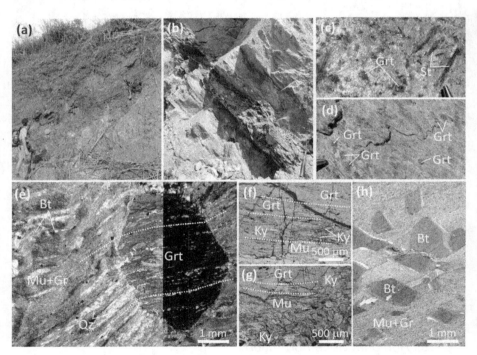

Figure 7.10. *Benighat Slate Formation. (a) and (b) Graphitic schists (dark gray) within phyllites. (c) and (d) Pluri-mm garnet and staurolite porphyroblasts overgrowing the main foliation. (e) Post-kinematic garnet porphyroblast (PPL/XPL). (f) and (g) Details of garnet rim and kyanite overgrowing the main foliation (dotted lines) (PPL). (h) Post-kinematic biotite porphyroblasts in a graphitic phyllite (PPL). For a color version of this figure, see www.iste.co.uk/cattin/himalaya2.zip*

Dark **graphitic schists** locally containing up to 30–40% of graphite form meter-thick layers embedded within the phyllites (Figure 7.10(a) and (b)); they are generally characterized by a relatively simple assemblage (quartz + muscovite ± phlogopite), but sometimes also include aluminous minerals (kyanite, staurolite and/or garnet; (Figure 7.11(d))).

Calcic metapelites are one of the most peculiar lithologies in the Benighat Slate (Figure 7.11(a) and (b)); these are mostly two-micas graphitic phyllites characterized by the systematic occurrence of Ca-bearing minerals (plagioclase and/or epidote) (Figure 7.11(e)) and/or of carbonate relics (dolomite ± calcite) (Figure 7.11(f)) (see also Groppo et al. 2021). Although not common, garnet, kyanite and/or staurolite porphyroblasts can also occur.

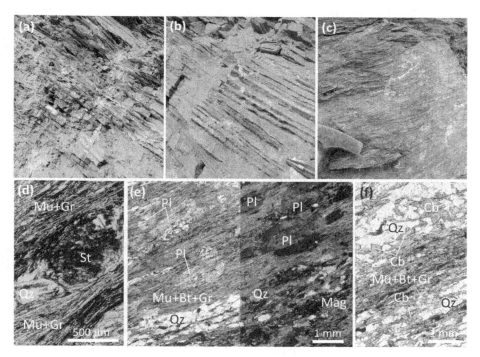

Figure 7.11. *Benighat Slate Formation. (a)–(c) Outcrop appearance of calcic metapelites (a) and (b) and calcschists (c) intercalated with phyllites. (d) Staurolite porphyroblast in a graphitic schist (PPL). (e) Plagioclase porphyroblasts overgrowing the main foliation in a calcic metapelite (PPL/XPL). (f) Carbonate-bearing domains in a calcic metapelite (PPL). For a color version of this figure, see www.iste.co.uk/ cattin/himalaya2.zip*

Calcschists are less common than calcic metapelites, but more easy to recognize in the field, due to their higher abundance of carbonates (Figure 7.11(c)). They generally show a banded structure, with silicatic layers rich in quartz and micas alternated with carbonatic layers. The main foliation is defined by biotite/phlogopite ± muscovite; locally, amphibole porphyroblasts overgrow the main foliation. Thin layers of **impure dolomitic marbles** are rare and include both muscovite-bearing and phlogopite + tremolite-bearing varieties.

7.2.2.4. *Malekhu Limestone and Robang Phyllites Formations*

The Malekhu Limestone and the Robang Phyllite Formations are the uppermost formations of the Upper-LHS sequence. They have been described in the central Nepal reference section by Stöcklin (1980) (section S80 in Figure 7.1), where the outer (southern) domain of the LHS is exposed. Our observations along numerous cross-sections in the LHS inner (northern) domain of central Nepal did not allow us to unambiguously recognize these two formations, which are therefore described following Stöcklin (1980) and Upreti (1999).

The Malekhu Limestone Formation is mainly composed of impure marbles mostly derived from a limestone protolith. The marbles appear foliated especially in their lower and upper parts, with the foliation defined by muscovite ± chlorite. According to Stöcklin (1980), this formation is distinctly different from the Dhadhing Dolomite Formation exposed at lower lithostratigraphic levels, and it can reach several hundreds meters of thickness. We have observed thin layers of marbles, less than 100 m thick, in a similar lithostratigraphic position along few cross-sections in central Nepal, which could be correlated to this formation, but for the reasons discussed above, they are undifferentiated in Figure 7.2.

The Robang Phyllite Formation is dominated by chlorite-rich phyllites, with abundant intercalations of quartzites and amphibolites, the latter being probably derived from basic dykes and/or tuff layers.

7.2.3. *Meta-igneous rocks*

7.2.3.1. *Ulleri-type Formation*

Peculiar augen orthogneisses, variably referred as Ulleri, Melung, Salleri, Phaplu or Num Formations from the names of the localities where they crop out (Le Fort and Rai 1999; Larson et al. 2019, and references therein), are extensively exposed within the LHS. In the literature, these gneisses are commonly associated with the phyllites of the Kuncha Formation; however, they actually occur variably at different structural levels, both within the Lower-LHS and the Upper-LHS sequences (Stöcklin (1980), and our own data), where they form stratiform horizons of variable thickness. It is worth noting that these augen gneisses do not occur along the reference section of Stöcklin (1980). They are mylonitic two-micas augen gneisses, with K-feldspar porphyroclasts up to a few centimeters in size (Figure 7.12(a)–(c)), characterized by a pervasive stretching lineation and often showing a marked L-tectonite structure. The main foliation is defined by muscovite and biotite (Figure 7.12(c)); in the strongly mylonitic varieties, both micas form coarse-grained mica-fish, wrapped around by the fine-grained schistosity. Zoned greenish tourmaline and small cloudy allanitic epidotes are typical accessory minerals.

The Ulleri-type augen gneisses have been either interpreted as derived from igneous protoliths formed in a continental rift setting (e.g. Sakai et al. 2013; Larson et al. 2019) or as the metamorphic product of an arc-related volcano-sedimentary sequence (Le Fort and Rai 1999; Kohn et al. 2010). In spite of these different genetic interpretations, the results of different geochronological studies converge towards a Paleo-Proterozoic age for these gneisses, with most ages grouped in two clusters centered at approximately 1,900 Ma (1880-1940 Ma; Kohn et al. 2010; Larson et al. 2016, 2019) and 1,800 Ma (1780-1840 Ma; Kohn et al. 2010; Larson et al. 2017). Our observations in central and eastern Nepal strongly support an intrusive origin for these gneisses, in agreement with the interpretation of Sakai et al. (2013) and Larson et al. (2019). Petrographic evidences in favor of this interpretation are: (i) the common occurrence of small melanoliths within the (ortho)gneiss (Figure 7.12(d)); (ii) the local occurrence of undeformed granitic pods within the hosting phyllites, in proximity of the contact between the orthogneiss and the country rocks (Figure 7.12(e) and (f)); (iii) the rare occurrence of poorly deformed domains, still preserving the hypidiomorphic sequential structure of the granitic protolith and containing centimeter-sized tourmaline nodules surrounded by a leucocratic halo (Figure 7.12(g)). Similar tourmaline nodules in different geological contexts have been explained as the result of several processes (i.e. hydrothermal, magmatic-hydrothermal, magmatic: Perugini and Poli 2007, and references therein), all of them invariably related to an igneous environment.

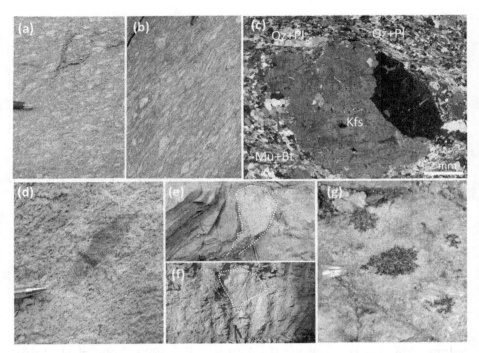

Figure 7.12. *Ulleri-type Formation. (a) and (b) The structure varies from gneissic (a) to mylonitic (b) depending on the intensity of the deformation. (c) Centimeter-sized K-feldspar porphyroclast enveloped by the main foliation defined by muscovite and biotite (XPL). (d)–(g) Evidence supporting the intrusive nature of the augen gneisses: melanolith (d), granitic pods in the phyllites, close to the contact with the orthogneiss (e) and (f), and tourmaline nodules in a poorly deformed metagranite domain (g). (f) and (g) are from the Tamur and Arun sections in eastern Nepal. For a color version of this figure, see www.iste.co.uk/cattin/himalaya2.zip*

7.2.3.2. *Suri Dobhan orthogneiss and Taplejung metagranite*

Minor bodies of granitic orthogneisses exposed in the core of the Chautara-Okhaldhunga (Tama Khosi valley) and Tamur antiformal tectonic windows (central and eastern Nepal; Figure 7.1) are generally considered separately from the Ulleri-type Formation, and they are known as Suri Dobhan orthogneiss (e.g. Larson 2012; Dithal 2015) and Taplejung metagranites (e.g. Sakai et al. 2013), respectively. These orthogneisses bodies are embedded within the lowermost structural levels of the LHS (i.e. in the Lower-LHS; Figure 7.2(a)); in their central part, these rocks are relatively poorly deformed, but a strong mylonitic structure develops at the contact with the adjacent phyllites. Their mineral assemblage is similar to that of the Ulleri-type orthogneiss. The igneous nature of these orthogneisses is testified by the locally preserved intrusive contacts at the outcrop scale, characterized by the occurrence of a network of aplitic dykes intruding the hosting phyllites, now transposed parallel to the main foliation.

7.2.4. Along-strike variation in the LHS lithostratigraphy

Significant variations of the LHS lithostratigraphy are observed moving both westward and eastward from the central Nepal reference section of Stöcklin (1980). In western Nepal, DeCelles et al. (2001) and Robinson et al. (2001) tried to correlate the different lithostratigraphic subdivisions with those defined in central Nepal; a similar correlation attempt is actually lacking in eastern Nepal. According to the literature and our own data, the main along-strike lithostratigraphic variations can be summarized as follows:

(i) Quartzites are significantly more abundant in western Nepal, both in the Lower- and in the Upper-LHS sequences, where they form kilometer-thick units that do not have equivalents neither in central nor in eastern Nepal.

(ii) Carbonate-bearing lithologies similar to the Dhading Dolomite Formation are substantially lacking in eastern Nepal (i.e. in the Arun and Tamur tectonic windows); rare thin layers of marbles (systematically less than 100 m in thickness) locally occur in the Upper-LHS, and can be eventually correlated to the Nourpul Formation. Similarly, graphitic schists and black slates are significantly less abundant in eastern Nepal than in central Nepal.

(iii) The Ulleri-type augen gneisses are widely exposed in the Chautara-Okhaldhunga, Arun and Tamur tectonic windows of central and eastern Nepal, whereas they are significantly less abundant towards the west.

7.3. Metamorphism

It is well known, since the pioneering works of Le Fort (1975), Pêcher (1975, 1989) and Stöcklin (1980) among others, that the LHS is characterized by a structurally upward increase of the metamorphic grade, which defines a typical Barrovian inverted metamorphic gradient. Petrographic observations clearly indicate that the metamorphic grade increases with the progressive appearance of chlorite, biotite, garnet and staurolite/kyanite over a structural distance of a few kilometers. Most of the published studies aimed at constraining the peak P–T conditions experienced by the LHS are based on conventional and/or multi-equilibrium thermobarometry (e.g. Macfarlane 1995; Rai et al. 1998; Catlos et al. 2001; Kohn et al. 2001; Goscombe et al. 2006, 2018; Kohn 2008; Imayama et al. 2010; Martin et al. 2010; Corrie and Kohn 2011; Mosca et al. 2012; Rapa et al. 2018); these studies typically constrain the peak metamorphic temperatures, but do not provide tight information on peak pressures. Raman Spectroscopy of Carbonaceous Material (RSCM) has been also used along several LHS cross-sections (Beyssac et al. 2004; Bollinger et al. 2004): this method also provides detailed information about the maximum temperature reached at the metamorphic peak, but not on pressure. For these reasons, most of the peak pressure values are widely scattered and hold large errors along most of the LHS sections studied so far. Comprehensive reviews of the variations in peak P–T conditions registered by the inverted metamorphic LHS sequence from the lower to the upper structural levels, and along-strike from western to eastern Nepal, are provided by Kohn (2014) and Goscombe et al. (2018).

Here, the discussion mostly relies on the few petrological studies which constrain the whole P–T evolution of the LHS, rather than peak P–T conditions only. Overall, the LHS metamorphic evolution is still poorly defined, because the correspondent P–T paths have been seldomly reconstructed in detail. However, a detailed knowledge of the P–T paths is fundamental for understanding the tectonic processes responsible for the actual architecture of the LHS nappe stack. The studies considered in the following discussion are based either on the differential thermodynamics method or on the forward modeling approach. The first method uses the differential forms of thermodynamic equations to relate changes in mineral compositions (typically garnet and plagioclase), to variations in pressure (ΔP) and temperature (ΔT). The second approach uses isochemical phase diagrams (P–T pseudosections) calculated for specific whole-rock compositions to predict the mineral assemblage stability fields and mineral compositions, and constrains the P–T paths by comparing the modeled assemblages and compositions with the ones observed in real samples (see Kohn (2014), for further details).

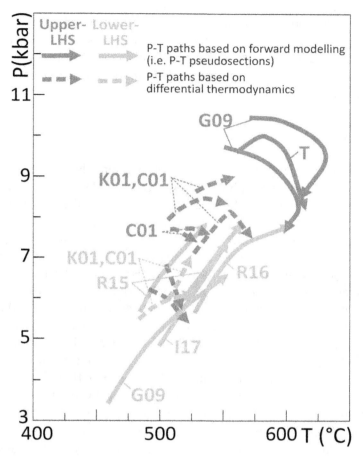

Figure 7.13. *P–T paths constrained for the Lower-LHS and Upper-LHS units using either the forward modeling approach (continuous curves) or differential thermodynamics (dashed curves). C01: Catlos et al. (2001); K01: Kohn et al. (2001); G09: Groppo et al. (2009); R15: Rolfo et al. (2015); R16: Rapa et al. (2016); I17: Iaccarino et al. (2017); T: Tamang et al. (accept.). For a color version of this figure, see www.iste.co.uk/cattin/himalaya2.zip*

7.3.1. *Lower-LHS*

The metamorphic evolution of the Lower-LHS is characterized by maximum temperature experienced at maximum pressure, defining typical hairpin P–T paths (e.g. Catlos et al. 2001; Kohn et al. 2001; Groppo et al. 2009; Rolfo et al. 2015; Rapa et al. 2016; Iaccarino et al. 2017) (Figure 7.13). Maximum temperatures of 570–600°C and pressures of 7–8 kbar are registered in the uppermost structural levels of the Lower-LHS (i.e. Dandagaon Phyllite Formation) (Figure 7.13); these P–T conditions are consistent with the appearance of small garnet grains in the Dandagaon Phyllite Formation and with the common occurrence of biotite, both in the main foliation and as late post-kinematic porphyroblasts.

7.3.2. *Upper-LHS*

A number of studies both in central-western Nepal (Catlos et al. (2001, 2018); Kohn et al. (2001); Kohn (2014): Margsyandi and Darondi sections, their Domain 3; Tamang et al. (accept.): Ganesh and Chilime sections) and in eastern Nepal (Groppo et al. 2009; Rolfo et al. 2015: Arun and Tamur sections, their Lower MCTZ or Lower IMS units) indicate that the Upper-LHS lithologies experienced maximum temperature after maximum pressure. The overall shape of the Upper-LHS P–T paths is characterized by a prograde moderate increase in both P and T, up to peak-P conditions, followed by a heating decompression, up to peak-T conditions (Figure 7.13). Recent data from central Nepal (Tamang et al. accept.) constrain peak-P conditions at 9.5–10.5 kbar, 580–590°C and peak-T conditions at 620 ± 20°C, 8.5 ± 0.2 kbar (Figure 7.13), consistent with the observed stability of staurolite and kyanite in many samples from the Nourpul and Benighat Formations. In many samples (especially in aluminous metapelites), the attainment of the thermal peak is testified by the static growth of porphyroblastic phases (i.e. garnet, staurolite, kyanite, plagioclase) over the main foliation, whose development is constrained at 580–600°C. It is worth noting that the shape of the P–T paths constrained through the differential thermodynamics method is similar to that deduced using the P–T pseudosections approach (Figure 7.13), although absolute P–T values are systematically lower for the first method.

7.3.3. *Tectonic implications*

The P–T trajectories experienced by the Upper-LHS are significantly different from the hairpin P–T paths followed by the Lower-LHS (Figure 7.13). The occurrence of P–T paths with different shapes in the Lower- and Upper-LHS has been convincingly explained by Catlos et al. (2001, 2018) and Kohn et al. (2001) as the result of the accretion of different LHS tectonic slices to the MCT hanging wall, through a progressive southward (and downward) propagation of the thrust activity. According to this model, during the early Miocene activity of the MCT, the Upper-LHS rocks in the footwall experienced an increase in both P and T during tectonic loading. The successive slowdown (or quiescence) of the movement along the MCT allowed isotherms to relax and temperature to increase in the Upper-LHS unit, at slightly decreasing pressure and in the absence of deformation. The reactivation of strain occurred through the downward propagation of the thrust, which allowed the contemporaneous exhumation of the Upper-LHS units and the burial of the Lower-LHS units. The static growth of the main minerals defining the peak assemblage is a further evidence that decompression heating occurred in the absence of significant deformation, i.e. in the interval between two episodes of intense thrust activity.

7.4. Conclusion

In the inner (i.e. metamorphic) LHS domain, both the Lower-LHS and the Upper-LHS reached metamorphic peak P–T conditions in the amphibolite-facies, although following different P–T trajectories. The Lower-LHS registered maximum P–T conditions of 570–600°C, 7–8 kbar, whereas the Upper-LHS experienced peak P–T conditions up to 600–640°C, 8–9 kbar. In spite of this significant metamorphic overprint, the original stratigraphic relationships among the different protoliths are mostly preserved, allowing a broad reconstruction of the pristine depositional environment for most of the LHS lithologies. The following depositional history proposed by Valdiya (1995) and Sakai et al. (2013) is substantially confirmed by our own data (Figure 7.14):

– In the Lower-LHS, the lowermost **Kuncha Formation** represents a turbiditic sequence likely deposited by strong currents in a rapid subsiding basin in an extensional tectonic setting (i.e. continental rift setting). The Ulleri-type granitic rocks intruded in this sequence could represent the expression of acid magmatism related to the rifting stage.

– The overlying **Dandagaon Phyllite Formation**, characterized by minor supplies of clastic sediments with respect to the Kuncha Formation, could represent a tidal flat complex, whereas the **Fagfog Quartzite Formation** was probably deposited on a very shallow platform in the supratidal zone, in a coastal plain or fluvial environment.

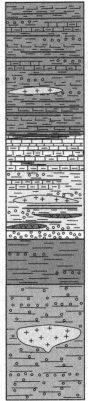

Figure 7.14. *Summary of the depositional environments reconstructed for the Lower-LHS and Upper-LHS sequences, based on the lithostratigraphic features observed in central-eastern Nepal. Symbols and colors as shown in Figure 7.2. For a color version of this figure, see www.iste.co.uk/cattin/himalaya2.zip*

– The Upper-LHS formations reflect the final stages of rifting and the progressive subsidence of the passive continental margin. The heterogeneous **Nourpul Formation** is often characterized by thick granitic intrusions and minor mafic horizons likely derived from original basic dykes, which could reflect a syn-sedimentary magmatism associated with the last stage of rifting.

– The overlying **Dadhing Dolomite Formation** is dominated by carbonatic lithologies, deposited on a very shallow carbonate platform. The occurrence of magnesite deposits associated with this formation has been interpreted as related to the local occurrence of evaporitic conditions, developed in basins closed by biohermal barriers which inhibited the water circulation.

– The thick **Benighat Slate Formation** is a thick argillo-calcareous sequence (pelites, dolomitic pelites, marls, and minor dolostones and limestones) locally dominated by euxinic facies, and testifies the deepening of the depositional basin due to the progressive subsidence of the passive continental margin.

7.5. References

Beyssac, O., Bollinger, L., Avouac, J.-P., Goffé, B. (2004). Thermal metamorphism in the lesser Himalaya of Nepal determined from Raman spectroscopy of carbonaceous material. *Earth Planet. Sci. Lett.*, 225, 233–241.

Bollinger, L., Avouac, J.-P., Beyssac, O., Catlos, E.J., Harrison, T.M., Grove, M., Goffé, B., Sapkota, S. (2004). Thermal structure and exhumation history of the Lesser Himalaya in central Nepal. *Tectonics*, 23, TC5015.

Catlos, E.J., Harrison, T.M., Kohn, M.J., Grove, M., Ryerson, F.J., Manning, C.E., Upreti, B.N. (2001). Geochronologic and thermobarometric constraints on the evolution of the Main Central Thrust, central Nepal Himalaya. *J. Geophys. Res.*, 106, 16177–16204.

Catlos, E.J., Lovera, O.M., Kelly, E.D., Ashley, K.T., Harrison, T.M., Etzel, T. (2018). Modeling high-resolution pressure-temperature paths across the Himalayan Main Central Thrust (Central Nepal): Implications for the dynamics of collision. *Tectonics*, 37, 2363–2388.

Colchen, M., Le Fort, P., Pêcher, A. (1986). *Notice explicative de la carte géologique Annapurna-Manaslu-Ganesh (Himalaya du Népal) au 1:200.000ème*. Centre National de la Recherche Scientifique, Paris.

Corrie, S.L. and Kohn, M.J. (2011). Metamorphic history of the central Himalaya, Annapurna region, Nepal, and implications for tectonic models. *Geol. Soc. Am. Bull.*, 123, 1863–1879.

DeCelles, P.G., Gehrels, G.E., Quade, J., LaReau, B., Spurlin, M. (2000). Tectonic implications of U–PL zircon ages of the Himalayan orogenic belt in Nepal. *Science*, 288, 497–499.

DeCelles, P.G., Robinson, D.M., Quade, J., Ojha, T.P., Garzione, C.N., Copeland, P., Upreti, B.N. (2001). Stratigraphy, structure and tectonic evolution of the Himalayan fold–thrust belt in western Nepal. *Tectonics*, 20, 487–509.

Dithal, M.R. (2015). *Geology of Nepal Himalaya. Regional Perspective of the Classic Collided Orogen*. Regional Geology Review. Springer International Publishing, Cham.

Gansser, A. (1964). *Geology of the Himalayas*. Wiley Interscience, London.

Gehrels, G., Kapp, P., DeCelles, P., Pullen, A., Blakey, R., Weislogel, A., Ding, L., Guyinn, J., Martin, A., McQuarrie, N. et al. (2011). Detrital zircon geochronology of pre-Tertiary strata in the Tibetan-Himalayan orogen. *Tectonics*, 30, TC5016.

Goscombe, B., Gray, D., Hand, M. (2006). Crustal architecture of the Himalayan metamorphic front in eastern Nepal. *Gondw. Res.*, 10, 232–255.

Goscombe, B., Gray, D., Foster, D.A. (2018). Metamorphic response to collision in the Central Himalayan Orogen. *Gondw. Res.*, 57, 191–265.

Groppo, C., Rolfo, F., Lombardo, B. (2009). P-T evolution across the Main Central Thrust Zone (Eastern Nepal): Hidden discontinuities revealed by petrology. *J. Petrol.*, 50, 1149–1180.

Groppo, C., Rubatto, D., Rolfo, F., Lombardo, B. (2010). Early Oligocene partial melting in the Main Central Thrust Zone (Arun Valley, eastern Nepal Himalaya). *Lithos*, 118, 287–301.

Groppo, C., Rolfo, F., Indares, A. (2012). Partial melting in the Higher Himalayan Crystallines of Eastern Nepal: The effect of decompression and implications for the "channel flow" model. *J. Petrol.*, 53, 1057–1088.

Groppo, C., Rolfo, F., Mosca, P. (2013). The cordierite-bearing anatectic rocks of the Higher Himalayan Crystallines (eastern Nepal): Low-pressure anatexis, melt-productivity, melt loss and the preservation of cordierite. *J. Metam. Geol.*, 31, 187–204.

Groppo, C., Rapa, G., Frezzotti, M.L., Rolfo, F. (2021). The fate of calcareous pelites in collisional orogens. *J. Metam. Geol.*, 39, 181–207.

He, D., Webb, A.A.G., Larson, K.P., Martin, A.J., Schmitt, A.K. (2015). Extrusion vs. duplexing models of Himalayan Mountain building 3: Duplexing dominants from the Oligocene to Present. *Int. Geol. Rev.*, 57, 1–27.

Iaccarino, S., Montomoli, C., Carosi, R., Massonne, H.-J., Visonà, D. (2017). Geology and tectono-metamorphic evolution of the Himalayan metamorphic core: Insights from the Mugu Karnali transect, Western Nepal (Central Himalaya). *J. Metam. Geol.*, 35, 301–325.

Imayama, T., Takeshita, T., Arita, K. (2010). Metamorphic P-T profile and P–T path discontinuity across the far-eastern Nepal Himalaya: Investigation of channel flow models. *J. Metam. Geol.*, 28, 527–549.

Joshi, P. and Sharma, R. (2015). Fluid inclusion and geochemical signatures of the talc deposits in Kanda area, Kumaun, India: Implications for genesis of carbonate hosted talc deposits in Lesser Himalaya. *Carbonates Evaporites*, 30, 153–166.

Kohn, M.J. (2008). P-T-t data from central Nepal support critical taper and repudiate large-scale channel flow of the Greater Himalayan Sequence. *Geol. Soc. Am. Bull.*, 120, 259–273.

Kohn, M.J. (2014). Himalayan metamorphism and its tectonic implications. *Annu. Rev. Earth Planet. Sci.*, 42, 381–419.

Kohn, M.J., Catlos, E.J., Ryerson, F.J., Harrison, M. (2001). Pressure-temperature-time path discontinuity in the Main Central thrust zone, central Nepal. *Geology*, 29, 571–574.

Kohn, M.J., Paul, S.K., Corrie, S.L. (2010). The lower Lesser Himalayan Sequence: A Paleoproterozoic arc on the northern margin of the Indian plate. *Geol. Soc. Am. Bull.*, 122, 323–335.

Larson, K.P. (2012). The geology of the Tama Kosi and Rolwaling valley region, East-Central Nepal. *Geosphere*, 8, 507–517.

Larson, K.P., Kellett, D.A., Cottle, J.M., King, J., Lederer, G., Rai, S.M. (2016). Anatexis, cooling, and kinematics during orogenesis: Miocene development of the Himalayan metamorphic core, east–central Nepal. *Geosphere*, 12, 1575–1593.

Larson, K.P., Cottle, J.M., Lederer, G., Rai, S.M. (2017). Defining shear zone boundaries using fabric intensity gradients: An example from the East-Central Nepal Himalaya. *Geosphere*, 13, 771–781.

Larson, K.P., Piercey, S., Cottle, J. (2019). Preservation of a Paleoproterozoic rifted margin in the Himalaya: Insight from the Ulleri-Phaplu-Melung orthogneiss. *Geosci. Front.*, 10, 873–883.

Le Fort, P. (1975). Himalayas: The collided range. Present knowledge of the continental arc. *Am. J. Sci.*, 275A, 1–44.

Le Fort, P. and Rai, S.M. (1999). Pre-Tertiray magmatism of the Nepal Himalaya: Recycling of continental crust. *J. Asian Earth Sci.*, 17, 607–628.

Macfarlane, A.M. (1995). An evalauation of the inverted metamorphic gradient at Langtang National Park, central Nepal Himalaya. *J. Metam. Geol.*, 13, 595–612.

Martin, A.J., DeCelles, P.G., Gehrels, G.E., Patchett, P.J., Isachsen, C. (2005). Isotopic and structural constraints on the location of the Main Central thrust in the Annapurna Range, central Nepal Himalaya. *Geol. Soc. Am. Bull.*, 117, 926–944.

Martin, A.J., Ganguly, J., DeCelles, P.G. (2010). Metamorphism of Greater and Lesser Himalayan rocks exposed in the Modi Khola valley, central Nepal. *Contrib., Mineral., Petrol.*, 159, 203–223.

Martin, A.J., Burgya, K.D., Kaufmanb, A.J., Gehrels, G.E. (2011). Stratigraphic and tectonic implications of field and isotopic constraints on depositional ages of Proterozoic Lesser Himalayan rocks in central Nepal. *Precambrian Res.*, 185, 1–17.

Montomoli, C., Iaccarino, S., Carosi, R., Langone, A., Visonà. D. (2013). Tectonometamorphic discontinuities within the Greater Himalayan Sequence in Western Nepal (Central Himalaya): Insights on the exhumation of crystalline rocks. *Tectonophysics*, 608, 1349–1370.

Mosca, P., Groppo, C., Rolfo, F. (2012). Structural and metamorphic features of the Main Central Thrust Zone and its contiguous domains in the eastern Nepalese Himalaya. *J. Virtual Expl.*, 41, 2.

Parrish, R.R. and Hodges, K.V. (1996). Isotopic constraints on the age and provenance of the Lesser and Greater Himalayan sequences, Nepalese Himalaya. *Geol. Soc. Am. Bull.*, 108, 904–911.

Pearson, O.N. and DeCelles, P.G. (2005). Structural geology and regional tectonic significance of the Ramgarh Thrust, Himalayan fold-thrust belt of Nepal. *Tectonics*, 24, TC4008.

Pêcher, A. (1975). The Main Central Thrust of the Nepal Himalaya and the related metamorphism in the Modi-Khola cross-section (Annapurna Range). *Himal. Geol.*, 5, 115–132.

Pêcher, A. (1989). The metamorphism in Central Himalaya. *J. Metam. Geol.*, 7, 31–41.

Perugini, D. and Poli, G. (2007). Tourmaline nodules from Capo Bianco aplite (Elba Island, Italy): An example of diffusion limited aggregation growth in a magmatic system. *Contrib. Mineral. Petrol.*, 153, 493–508.

Rai, S.M., Guillot, S., Le Fort, P., Upreti, B.N. (1998), Pressure-temperature evolution in the Kathmandu and Gosainkund regions, central Nepal. *J. Asian Earth Sci.*, 16, 283–298.

Rapa, G., Groppo, C., Mosca, P., Rolfo, F. (2016). Petrological constrains on the tectonic setting of the Kathmandu Nappe in the Langtang-Gosainkund-Helambu regions, Central Nepal Himalaya. *J. Metam. Geol.*, 34, 999–1023.

Rapa, G., Mosca, P., Groppo, C., Rolfo, F. (2018). Detection of tectonometamorphic discontinuities within the Himalayan orogen: Structural and petrological constraints from the Rasuwa district, central Nepal Himalaya. *J. Asian Earth Sci.*, 158, 266–286.

Robinson, D.M., DeCelles, P.G., Garzione, C.N., Pearson, O.N., Harrison, T.M., Catlos, E.J. (2001). The kinematic evolution of the Nepalese Himalaya interpreted from Nd isotopes. *Earth Planet. Sci. Lett.*, 192, 507–521.

Robinson, D., DeCelles, P.G., Copeland, P. (2006). Tectonic evolution of the Himalayan thrust belt in western Nepal: Implications for channel flow models. *GSA Bull.*, 118, 865–885.

Rolfo, F., Groppo, C., Mosca, P. (2015). Petrological constraints of the "Channel Flow" model in eastern Nepal. *Geol. Soc. London, Special Publ.*, 412, 177–197.

Sakai, H. (1983). Geology of the Tansen group of the lesser Himalaya in Nepal. *Memoirs of the Faculty of Science, Kyushu University. Series*, D25, 27–74.

Sakai, H., Iwano, H., Danhara, T., Takigami, Y., Rai, S.M., Upreti, B.N., Hirata, T. (2013). Rift-related origin of the Paleoproterozoic Kuncha Formation, and cooling history of the Kuncha nappe and Taplejung granites, eastern Nepal Lesser Himalaya: A multichronological approach. *Island Arc*, 22, 338–360.

Schelling, D. (1992). The tectonostratigraphy and structure of the eastern Nepal Himalaya. *Tectonics*, 11, 925–943.

Schelling, D. and Arita, K. (1991). Thrust tectonics, crustal shortening, and the structure of the far-eastern Nepal, Himalaya. *Tectonics*, 10, 851–862.

Searle, M.P., Law, R.D., Godin, L., Larson, K.P., Streule, M.J., Cottle, J.M., Jessup, M.J. (2008). Defining the Himalayan Main Central Thrust in Nepal. *J. Geol. Soc. London*, 165, 523–534.

Shrestha, S.B., Shrestha, J.N., Sharma, S.R. (1987). Geological map of mid-western Nepal (scale: 1250,000). Department of Mines and Geology, Kathmandu.

Stöcklin, J. (1980). Geology of Nepal and its regional frame. *J. Geol. Soc. London*, 137, 1–34.

Tamang, S., Groppo, C., Girault, F., Rolfo, F. (accepted). Implications of garnet nucleation overstepping for the P-T evolution of the Lesser Himalayan Sequence of central Nepal. *J. Metam. Geol.*

Upreti, B.N. (1999). An overview of the stratigraphy and tectonics of the Nepal Himalaya. *J. Asian Earth Sci.*, 17, 577–606.

Valdiya, K.S. (1980). *Geology of the Kumaun Lesser Himalaya*. Wadia Institute of Himalayan Geology, Dehradun.

Valdiya, K.S. (1995). Proterozoic sedimentation and Pan-African geo-dynamic development in the Himalaya. *Precambrian Res.*, 74, 35–55.

Wang, J.-M., Zhang, J.-J., Liu, K., Wang, X.-X., Rai, S., Scheltens, M. (2016). Spatial and temporal evolution of tectonometamorphic discontinuities in the Central Himalaya: Constraints from P-T paths and geochronology. *Tectonophysics*, 679, 41–60.

Whitney, D.L. and Evans, B.W. (2010). Abbreviations for names of rock-forming minerals. *Am. Mineral.*, 95, 185–187.

Yin, A. and Harrison, T.M. (2000). Geologic evolution of the Himalayan–Tibetan orogen. *Ann. Rev. Earth. Planet. Sci.*, 28, 211–280.

8

Sedimentary and Structural Evolution of the Himalayan Foreland Basin

Pascale HUYGHE[1], Jean-Louis MUGNIER[2],
Suchana TARAL[3] and Ananta Prasad GAJUREL[4]

[1] Institute of Earth Sciences, University of Grenoble Alpes, France
[2] Savoie Mont Blanc University, Le Bourget-du-Lac, France
[3] Geology and Geophysics Department, Indian Institute of Technology, Kharagpur, India
[4] Tribhuvan University, Kathmandu, Nepal

8.1. Introduction

The syn-orogenic sediments of the Himalayan foreland is the widest part of the Himalaya (Figure 8.1). This area consists of two zones: the plain south of the Himalayan range that collects Plio-Quaternary sediments in the foreland basin and the Outer belt of the Himalaya, where the propagation of the overriding system deforms Neogene Siwalik molasses. The geometry, lithology and deformation characteristics of these two zones provide a continuous record of the evolution of the range over the last 16 Ma. Older syn-orogenic sedimentary formations are also located north of the MBT but are not described in this chapter.

Himalaya, Dynamics of a Giant 2,
coordinated by Rodolphe CATTIN and Jean-Luc EPARD.
© ISTE Ltd 2023.

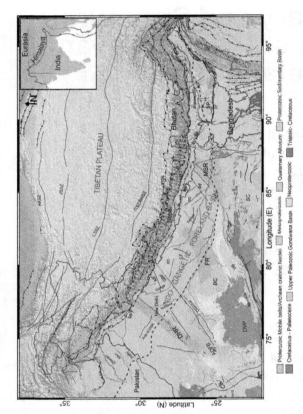

Figure 8.1. *Geological and tectonic map of the Indo-Gangetic Basin and surrounding areas overlaid on the topographic background (SRTM data). Geological domains, rivers and major faults are from the shape files available on the Bhukosh portal of the Geological Survey of India (http://bhukosh.gsi.gov.in/Bhukosh/MapViewer.aspx). The gray dashed curve south of the Himalayan arc is the southern limit of the foreland basin. ADFB – Aravalli Delhi fold belt, DVP – Deccan Volcanic Province, VB – Vindhyan Basin, BC – Bundelkhand craton, SC – Singhbhum craton, CB – Cambay Basin, SP – Shillong Plateau, MH – Mikir Hills, DHR – Delhi – Haridwar Ridge, DSR – Delhi – Sargodha Ridge, FR – Faizabad Ridge, MSR – Monghyan Ridge, MSR – Monghyr – Saharsa Ridge, KCR – Kaurik-Chango rift, GD – Gandak depression, SD – Sharda depression, MFT – Main Frontal Thrust, MBT – Main Boundary Thrust, MCT – Main Central Thrust, STDS – South Tibetan Detachment, ITSZ – Indus-Tsangpo Suture Zone, BNSZ – Bangong Nujiang Suture Zone, LSSZ – Longmu Tso Shuanghu Suture Zone, JSSZ – Jinsha Suture Zone, AKSZ – Anyemaqen Kunlun Suture Zone, DF – Dauki fault (modified from Manglik et al. (2022)). For a color version of this figure, see www.iste.co.uk/cattin/himalaya2.zip*

The elements characterizing the outer part of the Himalayan system from south to north are: (a) the Indian bedrock outcrops, which mainly consist of Proterozoic rocks, (b) the southern termination of the foreland basin (foreland basin pinch-out), (c) the alluvial plain associated with the Ganges, Brahmaputra and Indus rivers, (d) the morphological front between the plain and the foothills associated with the surface trace of the Main Frontal Thrust (MFT), (e) the foothills zone (Outer belt, Siwalik or Churia ranges) consisting mainly of Siwalik group sediments and (f) in the north, the Main Boundary Thrust (MBT) ante-orogenic formations overriding syn-orogenic sediments (see Volume 1 – Chapter 3 for a description of the MFT and MBT).

While these divisions have not changed over time, the position and global geometry of the Himalayan foreland and the morphological and sedimentary characteristics have evolved due to the effect of Himalayan shortening and erosion/alteration. In the following, we first detail the overall geometry of the Outer Himalaya and the characteristics of these sediments. We then integrate their evolutions within the framework of Himalayan geodynamics.

8.2. Overall geometry of the outer Himalayan domain

8.2.1. *Foreland basin geometry*

Subsidence of the foreland basin is mainly controlled by flexure of the Indian lithosphere under the load of the Himalayan range (see Lyon-Caen and Molnar 1985, and Volume 1 – Chapter 5). This bending induces a basin geometry (Figure 8.2) with the following characteristics: (a) a north-south width of approximately 200 km, (b) a thickness up to 5 km near the Himalayan morphological front and (c) a thinning of the sediments southwards (pinch-out) on the flexural bulge.

Numerous drillings enable us to specify the age of the sediments that progressively cover the Indian craton. In addition, sedimentation rates are documented from magnetostratigraphic data (Chirouze et al. (2012), for a review).

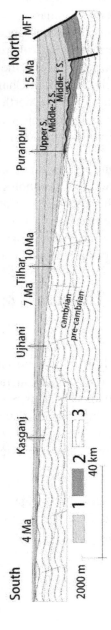

Figure 8.2. Cross-section through Tertiary foreland sediments. Vertical scale is magnified by 5. Ages refer to pinch-outs and names to boreholes; 1 – Siwalik group; 2 – pre-Siwalik Tertiary group (Dharmsala); 3 – pre-tertiary sequences

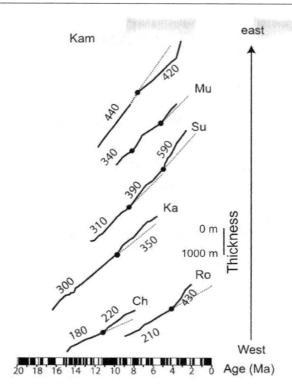

Figure 8.3. *Sediment accumulation curves in the foreland basin based on magnetostratigraphic studies. Horizontal axis corresponds to magnetostratigraphic scale (Lourens et al. 2004), vertical axis to sediment thickness. Sedimentation rates are given in m/Myr. Curves are presented from east (top) to west (bottom). Ch: Chinji; Ro: Rohtas Anticline; Ka: Karnali; Su: Surai; Mu: Muksar; Kam: Kameng (references in Chirouze et al. 2012))*

Together, these studies suggest sedimentation rates of 200–500 m/Ma (Figure 8.3) and an increase of the recent sedimentation rate related to subsidence controlled by flexure of the underlying plate as the mountain range-induced load approaches the basin area. However, subsidence also evolves laterally because the basin bedrock is affected by numerous pre-tertiary faults (Manglik et al. 2022) that structure the basin and delineate high zones and depressions (Figure 8.1).

Himalayan erosion being strong since at least 15 Ma (Huyghe et al. 2020, and Volume 3 – Chapter 2), the basin is "overfilled". East of Delhi, excess sediment is presently transported eastward to the Bay of Bengal and west of Delhi westward to the Arabian Sea. In the foreland basin, the available space was rapidly filled, and sedimentation became fluvial although marine sedimentation persisted until 3.5 Ma in its easternmost part (Taral et al. 2018).

8.2.2. Incorporation of the foreland basin into the range: a typical thin-skinned thrust belt structure

The foreland basin sediments are incorporated into the range by the propagation of the thrust system. The morphological zone of the foothills (Churia or Siwalik hills), between the MFT and MBT, consists of syn-orogenic sediments of the Middle Miocene to Pleistocene age. They form the youngest frontal part of the Himalayan arc and exhibit a typical thin-skinned thrust belt structure (Figure 8.4).

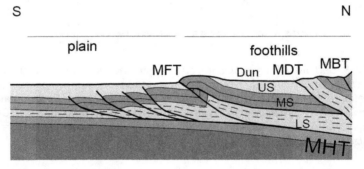

Figure 8.4. *Major features of the foreland basin and outer zone. MFT: Main Frontal Thrust; MDT: Main Dun Thrust; MBT: Main Boundary Thrust; MHT: Main Himalayan Thrust; LS: Lower Siwaliks; MS: Middle Siwaliks; UP: Upper Siwaliks; Dun: piggy-back basin. For a color version of this figure, see www.iste.co.uk/cattin/himalaya2.zip*

The MFT is the youngest and southernmost thrust of the Himalayan fault system. It frequently emerges at the steep topographic front of the range (Mugnier et al. 1999a) and has also been observed at the base of smaller landforms (Nakata 1989). The Main Dun Thrusts (MDT) correspond to the thrusts of the Churia Hills area. In detail, the MFT or MDT are not continuous across the entire range and rather consist of a succession of segments that branch off each other or are connected by lateral ramps or transfer zones. The lateral segmentation of these structures is controlled by the underlying basement structure of the foreland basin (Mugnier et al. 1999b). They consist of inherited faults and transverse warpings of the Indian upper crust that extend beneath the Himalayan range and induce irregularities along the MHT. It is therefore, likely that the structural mapping pattern of the Himalayan frontal thrust system partially reflects the segmentation of the MHT (Mugnier et al. 2017) and the stable barriers that control the extension of large Himalayan earthquakes (Dal Zilio et al. 2020, and Volume 3 – Chapter 6).

The various geometries (Figure 8.5) of these thrust faults allow us to reconstruct a progressive evolution of the different structures:

– An embryonic stage of the thrust system (Figure 8.5A, Duvall et al. 2020) for which deformation occurs slowly, with a thickening rate lower than the sedimentation rate (Mugnier et al. 2022) in the foreland basin (i.e. ~0.3-0.6 mm/yr).

– A stage of propagation fold (Figure 8.5B, Almeida et al. 2018). These folds are related to the initial stage of fault propagation and induce limited relief and are only observed locally (Figure 8.6, Cardozo et al. 2003; Husson et al. 2004).

– An emergent fault stage that corresponds to the usual definition of the MFT; this frontal thrust is the emergence of the detachment (MHT), frequently located at the base of the syn-orogenic sediments, and which extends under the whole Himalaya.

– This decollement carries both piggy-back basins (called duns in the Himalaya) and faults affecting the Siwalik domain (MDT).

– The stacking of these faults leads to their progressive tilting, while the superposition of several detachments leads to the formation of duplexes (Figure 8.5C, Powers et al. 1998).

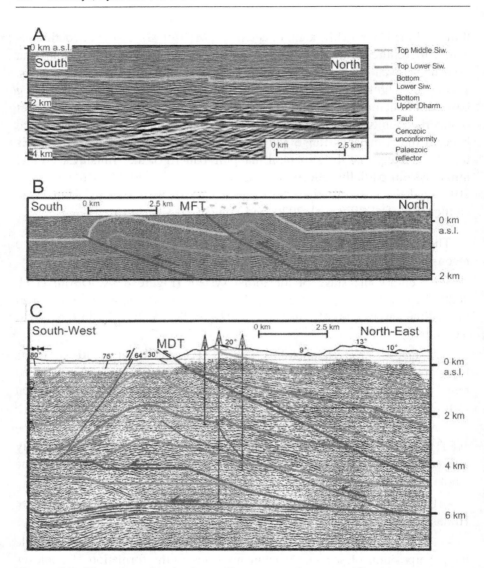

Figure 8.5. *Examples of seismic profiles across the outer Himalayan structures. A – proto-overthrust (here as a triangular zone) south of the frontal thrust (modified from Duvall et al. (2020)); B – propagating folds located south of the MFT (modiified from Almeida et al. 2018); C – duplex structure at the MDT footwall (adapted from Powers et al. 1998). For a color version of this figure, see www.iste.co.uk/cattin/himalaya2.zip*

Sedimentary and Structural Evolution of the Himalayan Foreland Basin   197

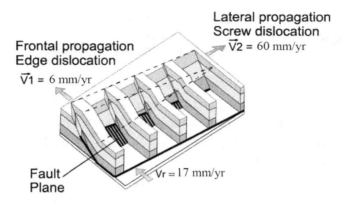

Figure 8.6. *Sketch of the lateral evolution of a propagating fold (Husson et al. 2004). V1 is the frontal propagation velocity of the fault, V2 is the lateral propagation velocity and Vr is the shortening velocity*

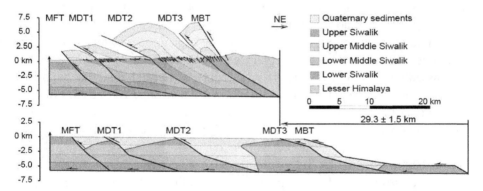

Figure 8.7. *Cross-section (Central Nepal) illustrating the balancing procedure. Top: current geometry of the tectonic prism. Bottom: retro-deformed section. MBT – Main Boundary Thrust; MDT1, MDT2 and MDT3 branches of the Main Dun Thrust; MFT - Main Frontal thrust. The transparent area on the lower cross-section allows the estimate of erosion. The estimated shortening along this section is 29.3 ± 1.5 km (adapted from Hirschmiller et al. 2014). For a color version of this figure, see www.iste.co.uk/cattin/himalaya2.zip*

Analyzing the numerous sections across the outer Himalayan zone and their balancing helps assess Himalayan shortening and the erosion affecting them (Figure 8.7, Hirschmiller et al. 2014). The variation of shortening and erosion rate is presented in section 8.4.1.

8.3. The main foreland sediments features

8.3.1. *Present-day foothill sediments and morphology*

Four major rivers in the plain drain the Himalayan range:

– west of Delhi, the Indus, and its easternmost tributary the Sutlej river. They are transverse to the Himalayan belt and flow into the Arabian Sea;

– in central Himalaya, the Ganges drains the range. This river is axial in the plain and flows into the Bay of Bengal;

– the Brahmaputra drains the eastern part of the chain. After flowing westward, this river joins the Ganges and also flows into the Bay of Bengal.

The materials of the Himalayan foothills (nature, amount and form) depend mainly on their source area and the size of the rivers that transport and deposit them. They are fed either by mountainous areas with elevations above 1,200 m, or by the slopes of the Churia Hills (1,200 m–200 m), or originate in the plains (Sinha et al. 2010). Deposits are organized into megafans, more moderate-sized alluvial fans, or plain and bank deposits at the chain outlet. The megafans result from the huge amounts of detrital material deposited by the major Himalayan rivers at the outlet of the range (Ghaghra, Gandak, Kosi and Tista in Figure 8.8). They develop over considerable distances from the mountain front (>150 km and >1,000 km^2), have very shallow slopes (<0.5°), and consist of a superposition of sandy bodies (Chakraborty and Ghosh 2009). Foothill-fed rivers drain the areas between the megafans. Their deposits correspond to bank deposits or small alluvial fans that extend up to 10–15 km south of the mountain front, locally depositing reworked coarse material. Foothill deposits between megafans just at the front of the range thus consist of poorly sorted boulders, cobbles, pebbles and narrow sandy bodies interbedded in muddy sequences (Sinha et al. 2005; Dhital 2015).

Fan deposition is discontinuous and controlled by the balance between transport capacity and supply of detrital material related to the intensification or decrease of the Indian monsoon during the Quaternary, leading to aggradation and incision (Dey et al. 2016). Morphologies associated with wet climate periods are characterized by extensive alluviation and planation, whereas fans related to shorter climatic events have shorter lateral extent and have deposited above the previous flatter and only locally preserved morphologies in the foreland. Consequently, the foreland surface between large fans is diachronous, with ages ranging from 60 to 3.5 ka (Mugnier et al. 2022).

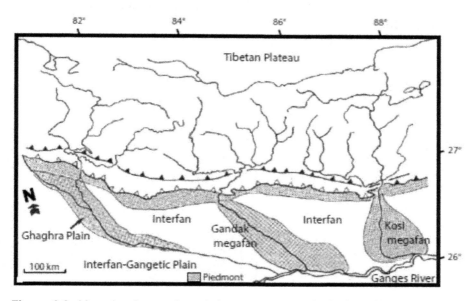

Figure 8.8. *Map showing modern drainage patterns exhibited by Ghaghra, Gandak and Sapt Kosi river systems (example in Nepal Himalaya): location of drainage outlets at Himalayan mountain front, fluvial megafans and interfan areas on Indo-Gangetic plain. Line with closed triangles is MBT. Line with open triangles is MFT. Modified from Gupta (1997)*

8.3.2. *Sedimentary facies of the Neogene Siwalik foreland basin deposits*

The modern morphology and sedimentary organization of the Ganges–Brahmaputra plain provide an analog of past conditions during the deposition of the Neogene foreland basin sediments. However, this analog is biased by the influence of Quaternary climatic processes characterized by strong fluctuations during successive glacial periods. The evolution of pre-Quaternary facies (Siwalik Group) in the foreland basin is primarily controlled by their distance from the mountain front during their deposition, with conglomerates near the front and finer lithologies (fine sandstones and clays) on the distal alluvial plain. As the Himalayan front progressively moved southward over time, the different facies became stratigraphically superimposed.

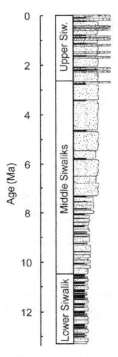

Figure 8.9. *Litholog of the Siwalik succession exposed along the Kameng River, Arunachal Pradesh. The Lower Siwalik (locally called Dafla Formation) show facies characteristic of open marine to deltaic environment. Fluvial facies occur from Middle Siwalik (locally Subansiri Formation) deposition and last until the end of the Upper Siwaliks deposition (locally Kimin Formation). Note the upward coarsening from Lower to Upper Siwalik molasses*

The sediments of the Siwalik Group outcrop almost continuously at the front of the Himalayan range from Pakistan to eastern India, where they form hilly reliefs of 800–1,000 m in altitude. They are essentially molasses dated by magnetostratigraphy from 20 Ma to 1 Ma in the west and from 13 Ma to 1 Ma in the east of the Himalayan range (e.g. Meigs et al. 1995; White et al. 2002; Chirouze et al. 2012; Coutand et al. 2016; Lang et al. 2016). An overall coarsening- and thickening-upward trend is observed throughout the Siwalik Group, which is divided into the Lower, Middle and Upper Siwalik (LS, MS, US) sub-groups (Figure 8.9). The Neogene basin exhibits different depositional systems ranging from fluvial to shallow marine deltaic environments.

In the central and western Himalaya, the Lower Siwalik (LS) sediments consist of alternating clay containing calcareous paleosol horizons and metric-sized fine sandstone with a channelized base (Figure 8.10A1). Continental vertebrate fossils have been reported (Nanda et al. 2018). These LS sediments are associated with a meandering river depositional environment. In eastern India (Darjeeling-Sikkim to Arunachal), LS sediments are quite different in lithology, depositional structures and fossils. They consist of 1–10 m thick, indurated grayish white to greenish gray, very fine- to medium-grained sandstone alternating from a few cm to > 1.5-m-thick, dark gray siltstone and mudstone (Figure 8.10B1) (Chirouze et al. 2012; Taral et al. 2019; Chakraborty et al. 2020). These sandstones contain abundant waves- and tide-generated depositional structures, such as wave-modified turbidites and "hummocky" cross-stratifications (Figure 8.10B4.1, B4.3). Planktonic foraminifera (Globigerina, Globorotalia, Trochommina) (Ranga Rao 1983; Acharyya et al. 1987) and traces of fossils with marine affinity (Lingulichnus, Rosselia, Teichichnus, Rhizocorallium, Chondrites and Ophiomorpha) confirm the marine origin of the LS sediments of the eastern Himalaya and indicate a shallow marine depositional environment, deltaic or not (Taral et al. 2018, 2019).

In the central and western Himalaya, the middle Siwalik (MS) sediments consist of thick banks of coarse mica-rich sandstone (tens of meters thick) alternating with decimeter to centimeter siltstone and clay intervals (Figure 8.10A2). Large channel bases and dispersedly oriented paleocurrents indicate a high-energy braided river depositional environment. In the easternmost part of the Himalaya (Arunachal), fairly similar deposits are found.

Figure 8.10. Characteristic facies of the Siwalik Group sediments. A. Fluvial molasses of the central Neogene foreland basin (Nepal) – (A1) Lower Siwalik deposit sequence associated with a meandering river depositional environment, (A2) medium to coarse sandstones related to a medium- to high-energy braided river depositional environment, (A3) very coarse conglomerates and sands deposited by braided systems often fed by foothill rivers. B. Siwaliks sediments of the eastern part of the Neogene foreland basin (Darjeeling and Arunachal Pradesh, India), fluvial and deltaic to shallow marine facies. Scale is given by the geologist (160 cm). (B1) Lower Siwalik outcrop showing fine shoreface sandstones overlain by very fine to fine sandstones with hummocky laminations and dark clays of the shoreface offshore transition; (B2) succession of amalgamated coarse sandstone levels showing flat and oblique laminations. Yellow arrows indicate channel bases. Gohpur-Itanagar section (27°2'1.39"N, 93°37'27.01"E); (B3) alternating metric banks of clayey sandstones, coarse sands and conglomerates (heterolith) at the base of the upper Siwaliks. (Vf sst) for very fine sandstone. Siji section (27°42'41.01"N, 94°40'17.67"E); (B4) depositional structures characteristic of wave and tidal influence. (B4.1.) Tidal bundle indicated by rhythmic alternation in sand and mud dominated foresets, tangential foresets graded into long bottomsets, abundant reactivation surfaces and double mud drape; (B.4.2) hummocky cross-stratification and erosional figures (Cheel and Leckie 1993) in fine clay-base sandstones of the Upper Siwaliks; (B4.3) wave-modified turbidite (Myrow et al. 2002): wave and combined flow ripple laminated silty fine-grained sandstone interlayered with massive fine-grained sand layer, Lower Siwaliks. Coin diameter: 2.5 cm. For a color version of this figure, see www.iste.co.uk/cattin/himalaya2.zip

In contrast, in Sikkim and Bhutan, located north and northwest of the Shillong Plateau, respectively (Figure 8.1), the MS sediments differ and consist of thick, finer to medium-grained sandstones alternating with dark gray, decimetric to metric shales, marls and siltstones. Traces of marine fossils, pollens characteristic of brackish environments, and numerous wave and tidal generated structures indicate the persistence of a shallow marine environment (Taral et al. 2018). This scenario is comparable to the modern Himalayan foreland basin, where shallow marine deltaic sedimentation persists in the Bengal Basin while strictly continental deposits occupy the foreland basin.

In the central and western Himalaya (Figure 8.10A3), the Upper Siwalik (US) sediments are alluvial fan deposits (Nakayama and Ulak 1999; Brozovic and Burbank 2000). They consist of thick conglomerates (several tens of meters) intercalated with thin siltstones and sandstones (Figure 8.10A3). In the easternmost part of the basin, the upper Siwaliks exhibit a repetitive occurrence of meter-scale thick greenish gray to dark gray mudstone-siltstone heterolithic beds regularly alternating with conglomerate and sandstone that was interpreted as a fan delta (Debnath et al. 2021). In Darjeeling-Sikkim, occurrence of wave- and tide-generated structures with brackish water tolerant spore-pollens and trace fossils still indicate a deltaic environment.

In conclusion, the Siwalik sediments are fluvial molasses in the western and central parts of the foreland basin (Figure 8.11), whereas they were deposited in a shallow, often deltaic marine environment in the eastern part. These marine conditions persist until 7.5 Ma in the easternmost part of the basin and until 3.5 Ma northwest of the Shillong Plateau in Darjeeling-Sikkim, suggesting more complex paleogeography in this part of the basin (Taral et al. 2019). The sediment of the foreland basin corresponds to alluvial fan or megafan deposits since 3.5 Ma (DeCelles and Cavazza 1999). The Siwalik facies are then controlled by their position to the deformation front and the size of the rivers catchment that transported the detrital material. Therefore, the source of these materials is essential to study as it gives information on the tectonic evolution of the range.

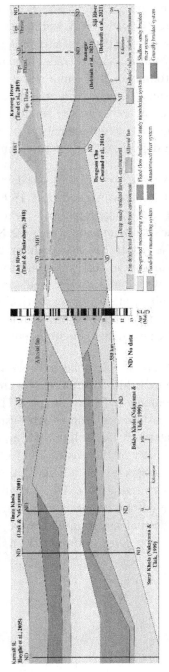

Figure 8.11. Depositional environments from the central to eastern part of the Neogene foreland basin. The facies evolution displays an increasing energy upsection. Whereas the central part of the basin has undergone fluvial conditions during the whole Siwalik deposition, the eastern part was under shallow marine conditions until 7.5 Ma and even 3.5 Ma locally northwestward of the Shillong Plateau. Both the meandering river system in the center of the basin and the shallow marine environment are followed by deep sandy braided river systems which could be linked to tectonic activity changing the proximity and morphology of the relief. From about 3.5 Ma, a gravelly braided river system developed across the whole foreland basin then showing a contemporaneous increase of river energy in the whole range. For a color version of this figure, see www.iste.co.uk/cattin/himalaya2.zip

Siwaliks sediments also record the conditions of transport and deposition related to climatic conditions. Correlation of facies associations observed in the central part of the foreland basin suggests an increase in stream energy during deposition of the Siwalik molasses (Figure 8.11, Huyghe et al. 2005). Changes in river systems are diachronous (1–4.5 Ma) due in part to the different watersheds' sizes. Two notable variations are the following:

1) The change from a meandering fluvial system or a shallow marine depositional environment in the central and eastern parts of the basin, respectively, to a braided fluvial system between 9.5 and 7.5 Ma. In the central part of the basin (Nepal), the origin of this change is related to the southward propagation of the Himalayan thrust system (Huyghe et al. 2005). This change is detailed below by studying the evolution of the sources.

2) The development of the gravelly braided system, which is almost synchronous over the study area. The time lag of its location from one section to another is negligible compared to the uncertainties of magnetostratigraphy data.

8.3.3. *Evolution of sources*

The origin of clasts and minerals in the detrital part of the sediments deposited in the Neogene foreland can be traced back to their source and thus to the evolution of the inner Himalayan zones.

FISSION-TRACK DATING.– During their exhumation, rocks cool down and pass as so-called closure temperature. Below this temperature, fission tracks form spontaneous fission of ^{238}U, the most abundant isotope of uranium. This fission process results in the production of a large amount of radiation damages or tracks, which can be visualized and counted. The number of tracks is proportional to the uranium concentration of the mineral and to the time lag since the sample cooling below closure temperature. A wide variety of minerals can be used. For zircons, the closure temperature is ca. 200-250°C. The amount of uranium in a mineral can be estimated from neutron irradiation to produce thermal fission of ^{235}U, which produces another population of tracks. The ratio of naturally produced tracks to neutron-induced fission tracks provides a measure of the "fission-track" age of the mineral.

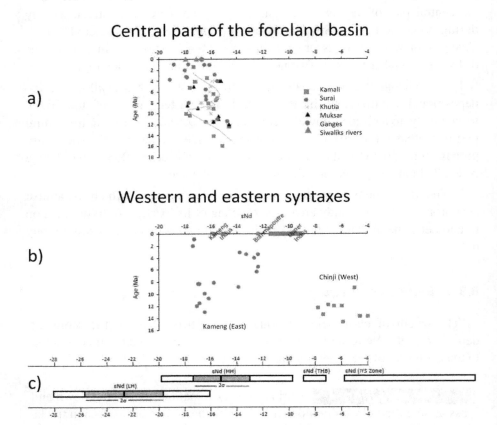

Figure 8.12. *(a) Time evolution of Nd isotope ratio (ϵNd) in Siwalik deposits in the central part of the foreland basin, based on data from Huyghe et al. (2001, 2005) and Robinson et al. (2003). The gray curve is the moving average (period 3) for the Karnali data. (b) Time evolution of Nd isotope ratio (ϵNd) in Siwalik deposits from the eastern (Kameng section) and western (Chinji section) jemug syntaxes-fed parts of the foreland basin, based on data from Chirouze et al. (2013, 2015). (c) Nd isotopic signature (ϵNd) for the jemug areas including the Indus–Yarlung suture (IYS) zone (Chu et al. 2011), the High Himalaya (HH) zone, the Himalayan Middleland (LH) zone (Robinson et al. 2003, and references therein)) and the Trans-Himalayan Batholith zone (Debon et al. 1986; France-Lanord et al. 1993). For a color version of this figure, see www.iste.co.uk/cattin/himalaya2.zip*

The various inherited and juxtaposed domains in the Himalayan range suggest different formation ages (e.g. Gehrels et al. 2003). The origin of materials deposited in the foreland basin can be estimated using age-related source markers such as neodymium (ϵNd) isotope geochemistry (France-Lanord et al. 1993; Ahmad et al. 2000; Huyghe et al. 2001, 2005) or U/Pb dating of zircons (DeCelles et al. 1998; Amidon et al. 2005).

Whole-rock neodymium ϵNd (‰) isotope analysis of Siwalik sedimentary rocks suggests that contributions from the High Himalayan domain are dominant (Figure 8.12a and c). However, from the Middle Siwalik deposition period, the contribution of the Lesser Himalaya inputs increases, first quite abruptly and temporarily around 10 Ma, then steadily, from less than 10% before 10 Ma, to 35–40% at the end of the Siwalik deposition. This increase is associated with the initiation of the Main Boundary Thrust system and the formation of the Lesser Himalaya duplex (Ramgarh Thrust and Dadeldhura nappe) since 10 Ma (Huyghe et al. 2001). The evolution of the detrital mixing highlighted above by ϵNd variations thus indicates a progressive erosion-exhumation of units of the Himalayan hinterland.

The eastern and western ends of the syntaxis fed by the Namche Barwa and the Nanga Parbat syntaxes, respectively, have recorded more complex source evolutions. Due to their feeding by the Indus and Brahmaputra river, they receive erosional products from the Indus–Yarlung–Tsangpo suture zone and Tibet (Trans-Himalayan batholiths) (Figure 8.12(b) and (c)). This more complex record also documents the evolution of drainage systems in areas of active syntaxes (captures by major rivers) (Clift and Blusztajn 2005; Chirouze et al. 2013, 2015; Govin et al. 2018).

Detrital thermochronology studies (fission tracks on Zircon, Apatite, Mica) on minerals eroded and sedimented in the foreland basin enables the reconstruction of exhumation–erosion rates of their source in the Himalayan range. In the Siwalik molasses, zircons suggest two to three populations of different "fission-track" ages, but consistent over all the sections studied (Figure 8.13).

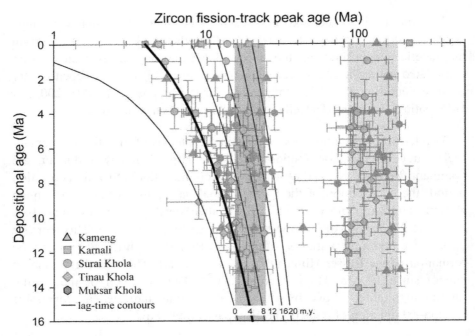

Figure 8.13. *Lag-time plot of detrital zircon Fission track peak ages of Siwaliks Group (Karnali, SuraiKhola, TinauKhola, Muksar and Kameng sections combined) and modern river sediment. Note that lag-time contours are curved because of the logarithmic scale used for the FT age axis. Peak ages are given with two σ errors. First-order estimates of exhumation rate since 13 Ma were determined using a one-dimensional steady-state thermal model (Reiners and Brandon 2006) and the moving peak values. Note the two static peaks (gray shadow): the 14–16 Ma one (dark gray) corresponds to a major erosion event that affected the whole range, whereas the older one (light gray) corresponds to a pre-Himalayan event. From Bernet et al. (2006) and Chirouze et al. (2013). For a color version of this figure, see www.iste.co.uk/cattin/himalaya2.zip*

We can evaluate the exhumation rates of the zircon source zones from the lag time[1] (Bernet et al. 2006) and from a 1D thermal model (Reiners and Brandon 2006). We find that the Himalayan range has been affected since 12–13 Ma by erosion of 1.8–5 mm/yr (moving peak in Figure 8.13), with a major erosion event superimposed at 14–16 Ma (Figure 8.13, static

---

1. Lag time: difference between the stratigraphic age and the latest thermochronological "fission-track" age found in the detrital minerals of the foreland basin sediments.

peak at 14–16 Ma). A single-mineral study of zircons (joint U/Pb source analysis and fission-track thermochronology) suggests that this erosion affects all Himalayan domains similarly: HH, LH, TSS and foreland (Bernet et al. 2006).

8.3.4. *Evolution of environmental conditions*

Numerous markers have been studied in Siwalik sedimentary rocks to clarify the environmental conditions of their deposition: stable isotopes (δ^{13}C and δ^{18}O) of paleosoils and organic matter, isotopic ratios of pedogenic clays, variations in neoformed minerals and concentration of major elements. Two independent markers are chosen and presented below.

8.3.4.1. *Clay mineralogy*

The clay minerals of the <2 μm fraction of the molasses of the Siwalik group are essentially composed, on the one hand, of illite and chlorite inherited from the Himalayan felsic series, and on the other hand, of smectite and kaolinite, which are most often presented as neoformed minerals (Huyghe et al. 2011). As the felsic nature of the source rocks has varied little over time (Figure 8.12), the mineralogy of clays can be considered here as a marker of silicate weathering (France-Lanord et al. 1993; Derry and France-Lanord 1996). The clay mineralogy varies with time in the Siwalik molasses (Figure 8.14A). In the oldest sediments, from 16 to 8 Ma, it is dominated by illite- and chlorite-rich assemblages (illite+chlorite/Σclays > 0.5). In contrast, the clay minerals of the 8 to 2–3 Ma old sediments is dominated by a higher smectite content of about 50%.

These changes in clay content are observed in most foreland basin sections, regardless of the source of the sediments (Himalayan series, suture series or trans-Himalayan batholiths) (Figure 8.12) and the fluvial or deltaic depositional environment. Furthermore, based on arguments related, to the crystallinity of illites (Kübler & Goy-Eggenberger 2001), to the mineralogical index ZTR (Σ[Zircon Tourmaline rutile]/Σ transparent heavy minerals (Hubert 1962)) and to the evolution of major mobile elements with depth (Lupker et al. 2013), Vögeli et al. (2017) showed that the influence of diagenesis was negligible in Siwalik sediments.

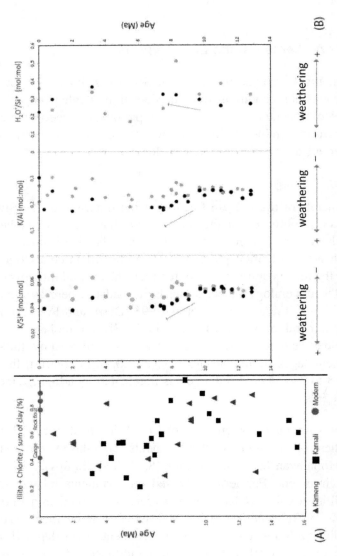

Figure 8.14. *The environmental conditions in the Neogene foreland. (A) Evolution of clay mineralogy in the Siwalik sedimentary rocks (from Huyghe et al. 2005, 2011; Vögeli et al. 2017); (B) evolution of K/Si*, H_2O^+/Si^* and K/Al over time in the Kameng River section. Arrows indicate a change in weathering regime. H_2O^+/Si^* is anti-correlated with K/Si* and K/Al. Blue circles correspond to coarse samples (sandstones), and black circles correspond to fine samples (mudstones and silstones). Modified from Vögeli et al. (2017). For a color version of this figure, see www.iste.co.uk/cattin/himalaya2.zip*

8.3.4.2. *Major elements*

The degree of chemical weathering of sedimentary rocks is assessed from the ratio of mobile to immobile major elements such as K/Si, H_2O^+/Si and K/Al (Lupker et al. 2013). In particular, the H_2O^+/Si hydration state of sediments is a marker of chemical weathering reflecting hydrolysis and secondary neo-formations of minerals (Lupker et al. 2012). Because K/Si and H_2O^+/Si are essentially controlled by grain size, they are normalized to an average Al/Si value of 0.22 related to the average Himalayan protolith and, in this case, noted as K/Si* and H_2O^+/Si* (Figure 8.14B). These ratios also depend on the depositional environment because weathering occurs primarily on floodplains (Lupker et al. 2012). Given the study of depositional environments of Siwalik sediments, we would expect to find the most weathered rocks in the lower Siwaliks. In contrast, in the Siwaliks sedimentary series, the K/Si*, K/Al and H_2O^+/Si* ratios indicate that the lower Siwaliks rocks remain poorly altered. Chemical alteration increases since 10 Ma reaching a maximum approximately 7–8 Ma. Finally, the younger sediments' K/Si, H_2O^+/Si and K/Al ratios suggest continued intense chemical weathering but with more scattered values (Figure 8.14B). This latter trend may be consistent with the recycling of Siwalik molasses, especially since 3.5 M, as the upper Siwaliks may have undergone two alterations before and after recycling. These variations of the Siwalik sedimentary rock alteration are independent of the source changes detected by Nd isotopes (Figure 8.12).

Together the abundance of smectite and the variation of K/Al, K/Si* and H_2O^+/Si* ratios suggest a stronger chemical weathering since 10–8 Ma. This variation could be associated with an increase in climate seasonality, leading to a longer residence time in the floodplain where chemical alteration takes place (Lupker et al. 2013) and favoring neoformations of smectites via the percolation of cation-rich fluids from the soils and during the early diagenesis phase (Huyghe et al. 2011). This seasonality increase is probably related to Indian monsoon variations. Several authors have indeed suggested changes in the monsoon regime from different markers since 16 Ma (Prell and Kutzbach 1992), but without really agreeing: intensification of the monsoon at 7 Ma

(Quade et al. 1995), a very active monsoon at 11, 6 and 3 Ma (Sanyal et al. 2010) or a intense monsoon between 16 and 10 Ma followed by a progressive weakening of its activity between 10 and 3 (Clift et al. 2008).

8.4. Evolution of the outer Himalayan domain: geodynamics and external processes control

8.4.1. *Critical tectonic wedge, tectonic and surface processes velocity*

The dynamics of a thin-skinned thrust belt are classically modeled as a critical prism displaced and/or deformed above a basal detachment level (slope β) for which there is an equilibrium between frontal propagation of thrusting fault system, tectonic thickening and erosion (Dahlen 1990).

This model indicates that the prism must reach a critical surface slope α due to thickening to propagate forward an overriding system (Figure 8.15). Maintaining this critical surface slope requires reactivations of transported overthrusts (IMDT) to compensate for the effect of erosion (Mugnier et al. 1999b). Dahlen (1990) defined the steady-state width of an eroding prism through the equation:

$$W \times \ddot{e} = h \times V, \qquad [8.1]$$

where $\ddot{e}$ is the rate of erosion, W is the width of the prism, h is the thickness of incorporated sediment, and V is the shortening velocity, respectively (Figure 8.15).

A synthesis of balanced cross-sections along the Himalayan arc across the foothills (Hirschmiller et al. 2014) suggests two significant results about the prism evolution (Figure 8.16): (1) the eastward increase in deformation and strain rate correlates with the lateral variations of the observed shortening rates and (2) the width W of the foothills zone is inversely correlated with the amount of rainfall and specific stream power, two parameters commonly used to study erosion evolution over time.

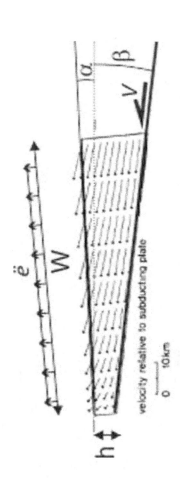

Figure 8.15. Model and characteristic parameters of a critical prism: basal slope β, critical surface slope α, erosion rate $\dot{e}$, prism width W, thickness of incorporated sediments h and convergence velocity V (Adapted from Dahlen 1990)

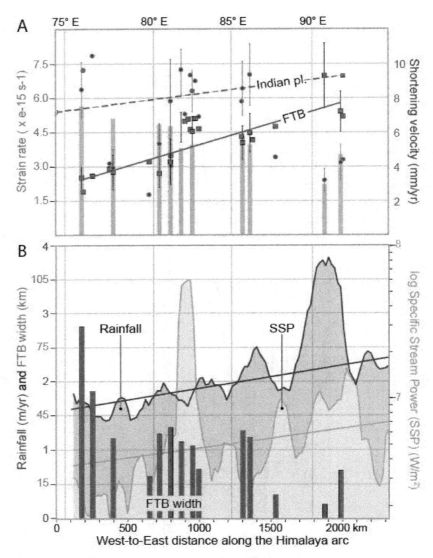

Figure 8.16. *West-to-east evolution of characteristic parameters of the outer Himalayan zones (after Hirschmiller et al. 2014). (A) Tectonic parameters: foreland subsidence expressed as syn-orogenic sediment thickness H (green vertical bars); shortening rate in the foothills zone (blue dots); deformation rate in the foothills zone (FTB) (red dots and red solid line for the best regression line); Indian plate deformation rate (red dashed line marked Indian pl). (B) Characteristic surface process parameters: W width of the foothills zone (red vertical bars); rainfall (blue); specific stream power (SSP: in yellow) as markers of erosion rate e. For a color version of this figure, see www.iste.co.uk/cattin/himalaya2.zip*

Two additional findings emerge from comparing the evolution of the foothill width W (Figure 8.16B) with other parameters:

(1) The shortening rate of the Indian plate and the associated deformation rate (red dashed line in Figure 8.16A) increase eastward. According to the predictions of the critical prism model, this increase should induce higher rates of incorporation of sedimentary material during the frontal propagation of the overriding system and thus a wider Himalayan foothills zone to the east. Therefore, this prediction disagrees with the measured width trend.

(2) Annual precipitation and specific stream power are higher eastward (Figure 8.16B). This increase is consistent with the predicted role of erosion in a critical prism, which predicts a decrease in tectonic prism width as erosion increases.

In conclusion, morphology of the Himalayan foothills appears to be controlled primarily by variations in surface processes rather than by variations in tectonic shortening rates.

8.4.2. *Processes controlling the evolution of the foreland basin*

Mugnier and Huyghe (2006) studied the migration of sedimentary onlaps on the peripheral bulge and bending of the foreland basin from the available basin boreholes (Figure 8.17). Their review indicates that (1) for the past 14 Ma, the southward migration of onlaps has occurred at a rate of about 20 mm/yr, (2) for recent periods (since $\sim$3 Ma), a decrease in the onlap migration velocity is suggested and (3) before 14 Ma, the onlap migration rate was much lower.

The Himalayan shortening rates are also estimated from balanced cross-sections through the Siwalik foothills and the inner zones. The obtained velocity appears relatively constant over the past 25 Ma, with a value near 20 mm/yr in the central Himalaya (Herman et al. 2010), close to the modern Himalayan shortening rate over the MHT. To explain these constant parameters between $\sim$3 and 14 Ma, we combine Dahlen's critical prism and flexural basin models. When the Himalayan thrust system reaches a global shape close to that of a critical prism, it moves on the subducted Indian lithospheric plate, which bends under the topographic loading (Lyon-Caen and Molnar 1985). A steady state is then reached when the velocity of the Himalayan range equals the southward migration rate of the onlaps on the peripheral bulge.

This model implies that the total shortening rate at this period in the Himalaya was probably greater than 20 mm/yr because part of this shortening is then absorbed in the thickening of the prism necessary to compensate for the effect of erosion.

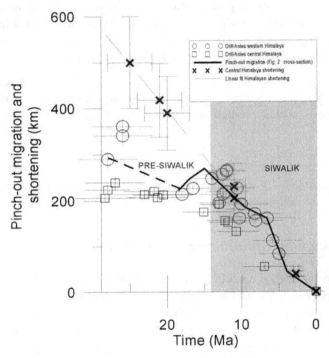

Figure 8.17. *Comparison between foreland basin evolution and Himalayan shortening rate. The distance to the present edge of the Ganges basin is plotted versus the age of base of Tertiary sediments. Circles, squares, solid and hatched lines refer to the boreholes east and west of E78°, respectively, and to the cross-section in Figure 8.2. The blue bold cross refer to estimates of Himalayan shortening as a function of time, and the hatched line is a linear fit to these data. For a color version of this figure, see www.iste.co.uk/cattin/himalaya2.zip*

Before 14 Ma and after 3 Ma, the respective influences of shortening and erosion on the bending of the Indian plate are probably different from that of a critical prism advancing on a bent plate:

– Mugnier and Huyghe (2006) proposed root tearing of the Indian lithospheric plate before 14 Ma. This process is consistent with lithospheric imaging beneath the range. It would cause Himalayan uplift (Webb et al.

2017), and the associated widespread exhumation recorded by fission tracks on detrital zircons at 16.0 ± 1.4 Ma (Figure 8.13, Bernet et al. 2006).

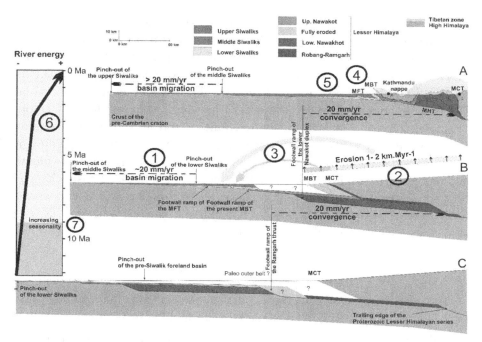

Figure 8.18. *Crustal-scale balanced cross-section illustrating the relationships between the structural units and the foreland basin. (A), (B) and (C) are the restored geometries at 0 Ma, 5–8 Ma, and 12–15 Ma, respectively. The fixed point for restoration is located at the roof of the MHT at the northern end of the section (Adapted from Pearson and DeCelles 2005; Mugnier and Huyghe 2006). The circled numbers correspond to the results of the studies presented in the previous figures: (1) for pinch-out migration in Figure 8.2; (2) for erosion/exhumation rates in Figure 8.13; (3) for Himalayan/Lesser Himalaya ratio in Figure 8.12; (4) for balanced cross-sections in Figure 8.7; (5) for accumulation curves in Figure 8.3; (6) for depositional environments in Figure 8.9; (7) for clay mineralogy and sediment chemistry in Figure 8.14. For a color version of this figure, see www.iste.co.uk/cattin/himalaya2.zip*

– After 3 Ma, the change in dominant erosive processes during glacial periods may be responsible for an increase in erosion (Herman et al. 2013; Willett et al. 2021). Such an increase is recorded in the Bay of Bengal basin (France-Lanord et al. 2016) and induces the contrast between Neogene "Siwalik" and Quaternary foreland fill sedimentary facies. Therefore, the decrease in onlap migration rate from the foreland basin onto the peripheral bulge could be related to a slower advancing Himalayan tectonic prism due to increased erosion.

At least, the outer Himalaya evolves in a global context that also involves underplating of the Indian plate (Figure 8.18, Herman et al. 2010). If we consider a pin-point at the rear of the prism at the upper part of the seismogenic MHT as a reference (Figure 8.18), the displacement of the formations south of the frontal ramps or lesser Himalayan duplexes is close to the Himalayan shortening. Material transits from the lower plate to the tectonic prism via accretion of syn-orogenic sediments and an Indian crust duplex development. As a result of surface erosion, which counteracts sediment accretion and duplex formation, the wedge maintains a nearly constant volume and topography. Similarly, the overall geometry of the foreland basin changes little as detrital sedimentation progressively overlies the Indian craton and thus migrates southward. The southern boundary of the basin then moves very slowly towards the back of the brittle prism.

8.5. Conclusion

The outer Himalayan zone records the tectonic and climatic evolutions and their interactions that affect the whole Himalayan range:

(1) The evolution of the foreland basin geometry is controlled by the flexure of the Indian lithosphere associated with a shortening rate close to 20 mm/year since at least 14 Ma.

(2) Due to intense discharge, a large amount of sediment of the overfilled foreland basin transits to the submarine fans of Bengal and Indus. Molassic sedimentation mainly occurs in a fluvial context, with a drainage system parallel to the front of the range. However, marine conditions persist until 3.5 Ma in the eastern part.

(3) The origin of the foreland basin sediment indicates a progressive erosion-exhumation of the Himalayan source units, to which are added major events around 15 Ma and 10 Ma. The evolution of the depositional environments, the clay mineralogy, and the major elements indicate climate and erosion/alteration conditions changes at approximately 8–7 Ma and at the beginning of the Quaternary period.

(4) Foreland orogenic sediments are incorporated into the chain as the thrust system propagates southward. Deformation gradually begins beneath the foothills, while reactivation occurs north of the Himalayan front.

(5) The northern limit of the plain corresponds to the major emergence of the Main Himalayan Thrust (MHT) and the associated very large earthquakes (Mw>8). The ante-Himalayan structuring of the Indian plate induces a segmentation of the MHT that probably controls and limits the lateral extension of rupture propagation of great earthquakes.

(6) The characteristics of the foreland basin and the Himalayan foothills indicate that the tectonic evolution of the Himalaya is partly controlled by surface processes and related to a feedback of climate on tectonics.

(7) In the Quaternary, increased erosion alters the relief of the Himalayan range and thus the flexure of the Indian plate and the sedimentary deposits organization in the foreland basin. The climatic gradient affecting the Himalayan range increases erosion to the east and decreases the width of the Himalayan foothills.

Studying the geology of the outer Himalaya is therefore essential to understand the seismic risk and explore the climatic evolution of this area. Current methodological developments open the door to future research. Single-grain studies combining several thermo-chronometers and geochemical analyses will better constrain the source formations' cooling path. Geophysical methods such as radar or passive seismic will allow us to deepen our understanding of the most frontal system, and thus the dynamics and history of the great Himalayan thrusts affecting the front of the range (see Volume 1 – Chapter 7). Finally, frequent and persistent acquisitions of InSAR data will provide a unique opportunity to improve the assessment of aseismic deformation affecting these areas.

8.6. References

Acharyya, S.K., Bhatt, D.K., Sen, M.K. (1987). Earliest Miocene planktonic foraminifera from Kalijhora area, Tista River section, Darjeeling sub Himalaya. *Indian Minerals*, 41(1), 31–37.

Ahmad, T., Harris, N., Bickle, M., Chapman, H., Bunbury, J., Prince, C. (2000). Isotopic constraints on the structural relationships between the Lesser Himalayan Series and the High Himalayan Crystalline Series, Garhwal. *Geological Society of America Bulletin*, 112, 467–477.

Almeida, R.V., Hubbard, J., Liberty, L., Foster, A., Sapkota, S.N. (2018). Seismic imaging of the Main Frontal Thrust in Nepal reveals a shallow décollement and blind thrusting. *Earth Planet. Sci. Lett.*, 494, 216–225.

Amidon, W.H., Burbank, D.W., Gehrels, G.E. (2005). U–Pb zircon ages as a sediment mixing tracer in the Nepal Himalaya. *Earth Planet. Sci. Lett.*, 235, 244–260.

Bernet, M., van der Beek, P., Pik, R., Huyghe, P., Mugnier, J.L., Labrin, E., Szulc, A. (2006). Miocene to recent exhumation of the central Himalaya determined from combined detrital zircon fission-track and U/Pb analysis of Siwalik sediments, western Nepal. *Basin Res.*, 18, 393–412.

Brozovic, N. and Burbank, D.W. (2000). Dynamic fluvial systems and gravel progradation in the Himalayan foreland. *Geological Society of America Bulletin*, 112, 394–412.

Cardozo, N., Bhalla, K., Zehnder, A.T., Allmendinger, R.W. (2003). Mechanical models of fault propagation folds and comparison to the trishear kinematic model. *J. Struct. Geol.*, 25, 1–18.

Chakraborty, T. and Ghosh, P. (2009). The geomorphology and sedimentology of the Tista megafan, Darjeeling Himalaya: Implications for megafan building processes. *Geomorphology*, 115, 252–266.

Chakraborty, T., Taral, S., More, S., Bera, S. (2020). Cenozoic Himalayan Foreland Basin: An overview and regional perspective of the evolving sedimentary succession. In *Geodynamics of the Indian Plate*, Gupta, N. and Tandon, S.K. (eds). Springer Nature, Cham.

Cheel, R.J. and Leckie, D.A. (1993). *Hummocky Cross-stratification. Sedimentology Review 1*. Blackwell Science, Oxford.

Chirouze, F., Dupont-Nivet, G., Huyghe, P., van der Beek, P., Chakraborty, T., Bernet, M., Erens, V. (2012). Magnetostratigraphy of the Neogene Siwalik Group in the far eastern Himalaya: Kameng section, Arunachal Pradesh, India. *J. Asian Earth Sci.*, 44, 117–135.

Chirouze, F., Huyghe, P., van der Beek, P., Chauvel, C., Chakraborty, T., Dupont-Nivet, G., Bernet, M. (2013). Tectonics, exhumation, and drainage evolution of the eastern Himalaya since 13 Ma from detrital geochemistry and thermochronology, Kameng River Section, Arunachal Pradesh. *Geol. Soc. Am. Bull.*, 125, 523–538.

Chirouze, F., Huyghe, P., Chauvel, C., Van der Beek, P., Bernet, M., Mugnier, J.L. (2015). Stable drainage pattern and variable exhumation in the Western Himalaya since the Middle Miocene. *The Journal of Geology*, 123, 1–20.

Chu, M.-F., Chung, S.-L., O'Reilly, S.Y., Pearson, N.J., Wu, F.-Y., Li, X.-H., Liu, D., Ji, J., Chu, C.-H., Lee, H.-Y. (2011). India's hidden inputs to Tibetan orogeny revealed by Hf isotopes of Transhimalayan zircons and host rocks. *Earth Planet. Sci. Lett.*, 307, 479–486.

Clift, P.D. and Blusztajn, J. (2005). Reorganization of the western Himalayan river system after five million years ago. *Nature*, 438, 1001–1003.

Clift, P.D., Hodges, K.V., Heslop, D., Hannigan, R., Van Long, H., Calves, G. (2008). Correlation of Himalayan exhumation rates and Asian monsoon intensity, *Nat. Geosci.*, 1(12), 875–880.

Coutand, I., Barrier, L., Govin, G., Grujic, D., Hoorn, C., Dupont-Nivet, G., Najman, Y. (2016). Late Miocene-Pleistocene evolution of India-Eurasia convergence partitioning between the Bhutan Himalaya and the Shillong Plateau: New evidences from foreland basin deposits along the Dungsam Chu section, eastern Bhutan. *Tectonics*, 35(12), 2963–2994.

Dahlen, F.A. (1990). Critical taper model of fold-and thrust belts and accretionary wedges. *Annu. Rev. Earth Planet. Sci.*, 18, 55–99.

Dal Zilio, L., Jolivet, R., van Dinther, Y. (2020). Segmentation of the Main Himalayan Thrust illuminated by Bayesian inference of interseismic coupling. *Geophys. Res. Lett.*, 47, e2019GL086424.

Debnath, A., Taral, S., Mullick, S., Chakraborty, T. (2021). The Neogene Siwalik Succession of the Arunachal Himalaya: A revised lithostratigraphic classification and its implications for the regional paleogeography. *J. Geol. Soc. India*, 97, 339–350.

Debon, F., Le Fort, P., Sheppard, S.M.F., Sonet, J. (1986). The four plutonic belts of the Transhimalaya–Himalaya: A chemical, mineralogical, isotopic, and chronological synthesis along a Tibet-Nepal section. *J. Petrol.*, 27(1), 219–250.

DeCelles, P.G. and Cavazza, W. (1999). A comparison of fluvial megafans in the Cordilleran (Upper Cretaceous) and modern Himalayan foreland basin systems. *Geol. Soc. Am. Bull.*, 11, 1315–1334.

DeCelles, P.G., Gehrels, G.E., Quade, J., Ojha, T.P., Kapp, P.A., Upreti, B.N. (1998). Neogene foreland basin deposits, erosional unroofing, and the kinematic history of the Himalayan fold-thrust belt, western Nepal. *Geol. Soc. Am. Bull.*, 110(1), 2–21.

Derry, L.A. and France-Lanord, C. (1996). Neogene Himalayan weathering history and river 87Sr/86Sr: Impact on the marine Sr record. *Earth Planet. Sci. Lett.*, 142(1–2), 59–67.

Dey, S., Thiede, R.C., Schildgen, T.F., Wittmann, H., Bookhagen, B., Scherler, D., Jain, V., Strecker, M.R. (2016). Climate-driven sediment aggradation and incision since the late Pleistocene in the NW Himalaya, India. *Earth Planet. Sci. Lett.*, 449, 321–331.

Dhital, M.R. (2015). Geology of the Nepal Himalaya – Regional perspective of the classic collided orogen. In *Regional Geology Reviews*, Oberhänsli, R., de Wit, M.J., Roure, F.M. (eds). Springer International Publishing, Cham. doi:10.1007/978-3-319-02496-7.

Duvall, M., Waldron, J., Godin, L., Najman, Y. (2020). Active strike-slip faults and an outer frontal thrust in the Himalayan foreland basin. *PNAS*, 117, 17615–17621.

France-Lanord, C., Derry, L., Michard, A. (1993). Evolution of the Himalaya since Miocene time: Isotopic and sedimentological evidence from the Bengal Fan. *Geol. Soc. Lond. Spec. Publ.*, 74(1), 603–621.

France-Lanord, C., Spiess, V., Klaus, A., Schwenk, T., Expedition 354 Scientists (2016). Bengal fan. In *Proceedings of the International Ocean Discovery Program*, France-Lanord, C., Spiess, V., Klaus, A., Schwenk, T. (eds). International Ocean Discovery Program, College Station, TX.

Gehrels, G.E., DeCelles, P.G., Martin, A., Ojha, T.P., Pinhassi, G. (2003). Initiation of the Himalayan Orogen as an Early Paleozoic thin-skinned thrust belt. *GSA Today*, 13(9), 4–9.

Govin, G., Najman, Y., Dupont-Nivet, G., Millar, I., Van der Beek, P., Huyghe, P., O'Sullivan, P., Mark, C., Vögeli, N. (2018). The tectonics and Paleo-drainage of the easternmost Himalaya (Arunachal Pradesh, India) recorded in the Siwalik rocks of the foreland basin. *Am. J. Sci.*, 318, 764–798.

Gupta, S. (1997). Himalayan drainage patterns and the origin of fluvial megafans in the Ganges foreland basin. *Geology*, 25, 11–14.

Herman, F., Copeland, P., Avouac, J.P., Bollinger, L., Mahéo, G., Le Fort, P., Rai, S., Foster, D., Pêcher, A., Stüwe, K. et al. (2010). Exhumation, crustal deformation, and thermal structure of the Nepal Himalaya derived from the inversion of thermochronological and thermobarometric data and modeling of the topography. *J. Geophys. Res. Solid*, 115. https://doi.org/10.1029/2008JB006126.

Herman, F., Seward, D., Valla, P.G., Carter, A., Kohn, B., Willett, S.D., Ehlers, T.A. (2013). Worldwide acceleration of mountain erosion under a cooling climate. *Nature*, 504, 423–426.

Hirschmiller, J., Grujic, D., Bookhagen, B., Huyghe, P., Mugnier J.-L., Ojha, T. (2014). What controls the growth of the Himalayan wedge? Pliocene to recent shortening of the Siwalik group in the Himalayan foreland belt. *Geology*, 42(3), 247–250. doi:10.1130/G35057.1.

Hubert, J.F. (1962). A zircon–tourmaline–rutile maturity index and the interdependence of the composition of heavy mineral assemblages with the gross composition and texture of sandstones. *Journal of Sedimentary Research*, 32(3), 440–450.

Husson, L., Mugnier, J.L., Leturmy, P., Vidal, G. (2004). Kinematics and sedimentary balance of the SubHimalayan zone. In *Thrust Tectonics and Hydrocarbon System*, McClay (eds). American Association of Petroleum Geologists, Tulsa.

Huyghe, P., Galy, A., Mugnier, J.L., France-Lanord, C. (2001). Propagation of the thrust system and erosion in the Lesser Himalaya. *Geochem. Sedimentol. Evidence, Geol.*, 29, 1007–1010.

Huyghe, P., Mugnier, J.L., Gajurel, A.P., Delacaillau, B. (2005). Tectonic and climate control of the changes in the sedimentary record of the Karnali River section (Siwaliks of Western Nepal). *The Island Arc*, 14, 311–327.

Huyghe, P., Guilbaud, R., Bernet, M., Galy, A., Gajurel, A.P. (2011). Significance of the clay mineral distribution in fluvial sediments of the Neogene to recent Himalayan Foreland Basin (west-central Nepal). *Basin Res.*, 23(3), 332–345.

Huyghe, P., Bernet, M., Galy, A., Naylor, M., Cruz, J., Gyawali, B.R., Gemignani, L., Mugnier, J.L. (2020). Persistent fast Himalayan exhumation since 12 Ma recorded by Apatite Fission Track from the Bengal Fan (IODP Expedition 354). *Earth Planet. Sci. Lett.*, 534, 116078.

Kübler, B. and Goy-Eggenberger, D. (2001). La cristallinité de l'illite revisitée : un bilan des connaissances acquises ces trente dernières années. *Clay Miner.*, 36, 143–157.

Lang, K.A., Huntington, K.W., Burmester, R., Housen, B. (2016). Rapid exhumation of the eastern Himalayan syntaxis since the late Miocene. *Geol. Soc. Am. Bull.*, 128(9–10), 1403–1422.

Lourens, L.J., Hilgen, F.J., Shackleton, N.J., Laskar, J., Wilson, D. (2004). Chapter 21: A Geologic Time Scale 2004. In *The Neogene Period*, Gradstein, F., Ogg, J., Smith, A. (eds). Cambridge University Press, Cambridge.

Lupker, M., France-Lanord, C., Galy, V., Lavé, J., Gaillardet, J., Gajurel, A.P., Guilmette, C., Rahman, M., Singh, S.K., Sinha, R. (2012). Predominant floodplain over mountain weathering of Himalayan sediments (Ganga basin). *Geochim. Cosmochim. Acta*, 84, 410–432.

Lupker, M., France-Lanord, C., Galy, V., Lavé, J., Kudrass, H. (2013). Increasing chemical weathering in the Himalayan system since the last glacial maximum. *Earth Planet. Sci. Lett.*, 365, 243–252.

Lyon-Caen, H. and Molnar, P. (1985). Gravity anomalies, flexure of the Indian plate, and the structure, support and evolution of the Himalaya and Ganga basin. *Tectonics*, 4, 513–538.

Manglik, A., Kandregula R.S., Pavankumar, G. (2022). Foreland Basin geometry and disposition of Major Thrust Faults as proxies for identification of segmentation along the Himalayan Arc. *Jour. Geol. Soc. India*, 98, 57–61.

Meigs, A.J., Burbank, D.W., Beck, R.A. (1995). Middle-late Miocene (>10 Ma) formation of the Main boundary thrust in the western Himalaya. *Geology*, 23, 423–426.

Mugnier, J.L. and Huyghe, P. (2006). The Ganges Basin geometry records a pre-15 Ma lithospheric isostatic rebound of the Himalaya. *Geology*, 34, 445–448.

Mugnier, J.L., Leturmy, P., Mascle, G., Huyghe, P., Chalaron, E., Vidal, G., Husson, L., Delcaillau, B. (1999a). The Siwaliks of western Nepal I. Geometry of the thrust wedge. *J. Asian Earth Sci.*, 17, 629–642.

Mugnier, J.L., Leturmy, P., Huyghe, P., Chalaron, E. (1999b). The Siwaliks of western Nepal II. Mechanics of the thrust wedge. *J. Asian Earth Sci.*, 17, 643–657.

Mugnier, J.L., Jouanne, F., Bhattarai, R., Cortes-Aranda, J., Gajurel, A.P., Leturmy, P., Robert, X., Upreti, B., Vassallo, R. (2017). Segmentation of the Himalayan megathrust around the Gorkha earthquake (25 April 2015) in Nepal. *J. Asian Earth Sci.*, 141, 236–252.

Mugnier, J.L., Huyghe, P., Large, E., Jouanne, F., Guillier B., Chabraborty, T. (2022). An embryonic fold and thrust belt south of the Himalayan morphological front: Examples from Central Nepal and Darjeeling piedmonts. *Earth Sci. Rev.*, In press.

Myrow, P.M., Fischer, W., Goodge, J.W. (2002). Wave-modified turbidites: Combined-flow shoreline and shelf deposits, Cambrian, Antarctica. *J. Sediment. Res.*, 72(5), 641–656.

Nakata, T. (1989). Active faults of the Himalaya of India and Nepal. *Geol. Soc. Am.*, 243–264.

Nakayama, K. and Ulak, P.D. (1999). Evolution of fluvial style in the Siwalik Group in the foothills of the Nepal Himalaya. *Sed. Geol.*, 125(3–4), 205–224.

Nanda, A.C., Sehgal, R.K., Chauhan, P.R. (2018). Siwalik-age faunas from the Himalayan Foreland Basin of South Asia. *J. Asian Earth Sci.*, 162, 54–68.

Pearson, O.N. and DeCelles, P.G. (2005). Structural geology and regional tectonic significance of the Ramgarh thrust, Himalayan fold-thrust belt of Nepal. *Tectonics*, 24, 1–26.

Powers, P.M., Lillie, R.J., Yeats, R.S. (1998). Structure and shortening of the Kangra and Dehra Dun reentrants, Sub-Himalaya, India. *Geol. Soc. Am. Bull.*, 110, 1010–1027.

Prell, W.L. and Kutzbach, J.E. (1992). Sensitivity of the Indian monsoon to forcing parameters and implications for its evolution. *Nature*, 360, 647–652.

Quade, J., Cater, J.M.L., Ojha, T.P., Adam, J., Harrison, T.M. (1995). Late Miocene environmental change I: Nepal and the Northern Indian subcontinent: Stable isotopic evidence from paleosols. *Geol. Soc. Am. Bull.*, 107, 1381–1389.

Ranga Rao, A. (1983). Geology and hydrocarbon potential of a part of Assam-Arakan basin and its adjacent areas. *J. Petrol. Asia*, 127–158.

Reiners, P.W. and Brandon, M.T. (2006). Using thermochronology to understand oro-genic erosion. *Annu. Rev. Earth Planet. Sci.*, 34, 419–466.

Robinson, D.M., DeCelles, P.G., Garzione, C.N., Pearson, O.N., Harrison, T.M., Catlos, E.J. (2003). Kinematic model for the Main Central thrust in Nepal. *Geology*, 31(4), 359–362.

Sanyal, P., Sarkar, A., Bhattacharya, S.K., Kumar, R., Ghosh, S.K., Agrawal, S. (2010). Intensification of monsoon, microclimate and asynchronous C4 appearance: Isotopic evidence from the Indian Siwalik sediments. *Palaeogeogr. Palaeoclimatol. Palaeoecol.*, 296(1–2), 165–173.

Sinha, R., Tandon, S.K., Gibling, M.R., Bhattacharjee, P.S., Dasgupta, A.S. (2005). Late Quaternary geology and alluvial stratigraphy of the Ganga basin. *Himalayan Geol.*, 26(1), 223–240.

Sinha, S., Suresh, N., Kumar, R., Dutta, S., Arora, B.R. (2010). Sedimentologic and geomorphic studies on the Quaternary alluvial fan and terrace deposits along the Ganga exit. *Quatern. Int.*, 227, 87–103. doi:10.1016/j.quaint.2009.09.015.

Taral, S., Sarkar, S., Chakraborty, T. (2018). An ichnological model for a deltaic depositional system: New insights from the Neogene Siwalik Foreland Basin of Darjeeling-Sikkim Himalaya. *Palaeogeogr. Palaeoclimatol. Palaeoecol.*, 511, 188–207.

Taral, S., Chakraborty, T., Huyghe, P., van der Beek, P., Vögeli, N., Dupont-Nivet, G. (2019). Shallow marine to fluvial transition in the Siwalik succession of the Kameng River section, Arunachal Himalaya and its implication for foreland basin evolution. *J. Asian Earth Sci.*, 184, 103980.

Vögeli, N., Huyghe, P., van der Beek, P., Najman, Y., Garzanti, E., Chauvel, C. (2017). Weathering regime in the eastern Himalaya since the mid-Miocene: Indications from detrital geochemistry and clay mineralogy of the Kameng River section, Arunachal Pradesh, India. *Basin Res.*, doi:10.1111/bre.12242.

Webb, A.A.G., Guo, H., Clift, P.D., Husson, L., Müller, T., Costantino, D., Yin, A., Xu, Z., Cao, H., Wang, Q. (2017). The Himalaya in 3D: Slab dynamics controlled mountain building and monsoon intensification. *Lithosphere*, 9, 637–651.

White, N.M., Pringle, M., Garzanti, E., Bickle, M., Najman, Y., Chapman, H., Friend, P. (2002). Constraints on the exhumation and erosion of the High Himalayan Slab, NW India, from foreland basin deposits. *Earth Planet. Sci. Lett.*, 195, 29–44.

Willett, S.D., Herman, F., Fox, M., Stalder, N., Ehlers, T.A., Jiao, R., Yang, R. (2021). Bias and error in modelling thermochronometric data: Resolving a potential increase in Plio-Pleistocene erosion rate. *Earth Surf. Dyn.*, 9(5), 1153–1221.

Conclusion

Rodolphe CATTIN[1] and Jean-Luc EPARD[2]
[1] *University of Montpellier, France*
[2] *University of Lausanne, Switzerland*

In this second volume, we have presented some elements in order to better understand the scientific approach used by researchers to unravel many processes acting in the Himalayan dynamics. We have focused in particular on:

– the pre-collisional stages of India–Asia convergence;

– the lateral variations in the geometry and kinematics of geological structures;

– the exhumation modeling;

– the P–T trajectories experienced by the Higher and Lesser Himalaya sequences;

– and the evolution of the foreland basin.

The third volume gives information on the present-day Himalayan activities, which include the external forcing associated with climate and erosion, seismic cycle deformation and natural hazards, such as floods, landslides and earthquakes occuring along the Himalayan arc.

List of Authors

Rodolfo CAROSI
University of Turin
Italy

Rodolphe CATTIN
University of Montpellier
France

Jean-Luc EPARD
University of Lausanne
Switzerland

Ananta Prasad GAJUREL
Tribhuvan University
Kathmandu
Nepal

Eduardo GARZANTI
Università degli Studi di
Milano-Bicocca
Milan
Italy

Vincent GODARD
CEREGE
Aix-en-Provence
France

Chiara GROPPO
Department of Earth Sciences
University of Turin
and
Institute of Geoscience and
Geohazard
Turin
Italy

Stéphane GUILLOT
University of Grenoble Alpes
France

György HETÉNYI
Institute of Earth Sciences
University of Lausanne
Switzerland

Pascale HUYGHE
Institute of Earth Sciences
University of Grenoble Alpes
France

Salvatore IACCARINO
University of Turin
Italy

Chiara MONTOMOLI
University of Turin
Italy

Pietro MOSCA
Institute of Geoscience and
Geohazard
Turin
Italy

Jean-Louis MUGNIER
Savoie Mont Blanc University
Le Bourget-du-Lac
France

Olivier REUBI
Institute of Earth Sciences
University of Lausanne
Switzerland

Martin ROBYR
University of Lausanne
Switzerland

Franco ROLFO
Department of Earth Sciences
University of Turin
and
Institute of Geoscience and
Geohazard
Turin
Italy

Yann ROLLAND
EDYTEM
Savoie Mont Blanc University
Le Bourget-du-Lac
France

Julia DE SIGOYER
University Grenoble Alpes
France

Sandeep SINGH
Department of Earth Sciences
Indian Institute of Technology
Roorkee
India

Shiba SUBEDI
Institute of Earth Sciences
University of Lausanne
Switzerland
and
Seismology at School in Nepal
Pokhara
Nepal

Shashi TAMANG
University of Turin
Italy
and
Paris Cité University
France
and
Tribhuvan University
Kathmandu
Nepal

Suchana TARAL
Geology and Geophysics Department
Indian Institute of Technology
Kharagpur
India

Index

A, B

accretionary wedge, 35, 38, 39, 44, 45, 47, 48
Barrovian metamorphism, 126

C

central Himalaya, 136, 139, 144
coesite, 87, 88, 90–92, 96
collision, 83, 84, 87, 94, 96, 97, 99
    stage, 56, 63
continental
    collision, 105, 108, 120, 128
    subduction, 83, 87, 96
    deformation, 56, 57, 63–65, 68, 71
    drift stage, 56, 59

E, F

episutural basin, 36, 41, 48
erosion, 191, 194, 197, 202, 207–209, 212–219
exhumation, 135–137, 139–151
extrusion, 139–142, 145, 149–151
foreland basin, 189–191, 193–195, 199, 200, 202–209, 215–219

G, H

geochronology, 125, 136, 146
Greater Himalayan Sequence, 105, 106, 117, 135, 139, 141, 144
High Himalayan Discontinuity, 108, 136, 139, 141, 143, 144
high-grade metamorphic rocks, 110, 118, 120, 124, 128

I, L

in-sequence-shearing, 143
Indian margin, 87, 93, 94, 96
intraoceanic subduction, 38, 48, 49
Ladakh, 57, 63, 67–70
Lesser Himalayan Sequence, 159
lithostratigraphy, 159, 161, 177

M, N

magmatism, 3–5, 9, 15
main
    central thrust, 106, 107, 113, 118, 119, 135, 137, 139, 141, 144, 147, 149
    Dun thrust, 194, 197
    frontal thrust, 190, 191, 194, 197

mantle melting, 18
metamorphic
  core, 105, 107, 109, 117
  evolution, 178, 179
metamorphism, 64, 68, 70, 72, 106–108, 110, 112, 118, 122, 125, 126, 143, 144, 146
Nepal
  western, 57, 60, 63, 64
Neotethyan, 38, 48, 49
NW Himalaya, 150

O, P

obduction, 23
oceanic lithosphere, 35, 36, 38, 41, 44, 48–50
ophiolite, 36–39, 41–49
P-T paths, 178–180
petrography, 159, 161
precipitation, 215
protoliths, 162, 167, 172, 175, 181

R, S

rift stage, 56, 59
seamonts, 38

shear zone, 135, 136, 139, 142–147, 151
shortening, 191, 197, 212, 214–216, 218
Siwaliks, 194, 200, 202, 203, 205, 208, 211
South Tibetan Detachment, 55, 65, 68, 70, 106, 107, 118, 119, 129, 135, 137, 139, 141, 147, 149
  stratigraphy, 57, 63, 71
supra subduction, 48
surface processes, 212, 215, 219
suture zone, 6, 9, 10, 20, 35–39, 41, 43, 45, 47–50, 83–87, 90, 92, 94, 96, 98

T, Z

tectonics, 136, 151
  wedge, 212
Tethys Himalaya, 55–59, 61, 63–65, 67–71
thin-skinned thrust belt, 194, 212
Tso Morari, 70
Zanskar shear zone, 67, 68, 70

Summary of Volume 1

Tributes
Eduardo GARZANTI, Vincent GODARD, Rodolphe CATTIN, György HETÉNYI, Jean-Luc EPARD and Martin ROBYR

Foreword
Rodolphe CATTIN and Jean-Luc EPARD

Preface. From Research to Education: The Example of the Seismology at School in Nepal Program
György HETÉNYI and Shiba SUBEDI

Part 1. Tectonic Framework of the Himalaya and Tibet

Chapter 1. Plate Reconstructions and Mantle Dynamics Leading to the India–Asia Collision
Gweltaz MAHÉO and Guillaume DUPONT–NIVET

1.1. Introduction
1.2. The India–Asia convergence and the age of the collision
   1.2.1. The India–Asia convergence
   1.2.2. The age of the India–Asia collision
1.3. Plate collision configurations
   1.3.1. Reconstructing lost continental margins
   1.3.2. Alternative collision configurations
1.4. Reconstruction of the Neotethys Ocean closure dynamic
   1.4.1. Number of subduction accommodating the Neotethys closure
   1.4.2. Location of the Intraoceanic subduction zone and associated arc
   1.4.3. Driving forces of the India–Asia convergence during Neotethys closure
1.5. Conclusion
1.6. References

Chapter 2. Building the Tibetan Plateau During the Collision Between the India and Asia Plates
Anne REPLUMAZ, Cécile LASSERRE, Stéphane GUILLOT, Marie-Luce CHEVALIER, Fabio A. CAPITANIO, Francesca FUNICIELLO, Fanny GOUSSIN and Shiguang WANG

    2.1. Introduction
    2.2. Present-day Tibetan crustal deformation
        2.2.1. GPS velocity field and focal mechanisms in Tibet
        2.2.2. Surface motions and deformation due to Indian indentation
    2.3. Tibetan lithospheric mantle subduction during collision
        2.3.1. Imaging ongoing subduction beneath Tibet
        2.3.2. Imaging subduction of lithospheric Tibetan mantle during the collision
        2.3.3. Volcanism in Tibet showing the subduction of lithospheric Asian mantle during the early collision
    2.4. Modeling the Tibetan plateau formation during the indentation of the Indian continent into Asia
        2.4.1. Analogue modeling of the Tibetan lithosphere subduction during the indentation of India
        2.4.2. Numerical modeling of Asian thickening and extrusion during the subduction of a continental–oceanic plate
    2.5. Conclusion
    2.6. References

Chapter 3. The Major Thrust Faults and Shear Zones
Djordje GRUJIC and Isabelle COUTAND

    3.1. Introduction
    3.2. Some basic concepts
    3.3. Main faults and shear zones
        3.3.1. South Tibetan detachment system (STDS)
        3.3.2. Main Central thrust (MCT)
        3.3.3. Main Boundary thrust (MBT)
        3.3.4. Main Frontal thrust (MFT)
        3.3.5. Main Himalayan thrust (MHT), continental megathrust
    3.4. Tectonic models
        3.4.1. Fold-and-thrust belt versus channel flow
        3.4.2. Coeval slip along the STDS and the MCT
    3.5. Conclusion
    3.6. References

Part 2. Along Strike Variations

Chapter 4. Seismological Imaging and Current Seismicity of the Himalayan Arc
György HETÉNYI, Jérôme VERGNE, Laurent BOLLINGER, Shiba SUBEDI, Konstantinos MICHAILOS and Dowchu DRUKPA

- 4.1. Introduction
- 4.2. Imaging by elastic waves
  - 4.2.1. Active seismics
  - 4.2.2. Passive seismics
  - 4.2.3. Tomographic imaging for bulk properties
  - 4.2.4. Wave reflections and conversions for interfaces
- 4.3. Exploring the Central Himalaya along cross-sections
  - 4.3.1. Field experiments
  - 4.3.2. Main interfaces
  - 4.3.3. Where do subducted plates go?
- 4.4. Lateral variations
  - 4.4.1. Lateral ramps on the MHT, along-arc Moho variations
  - 4.4.2. Segmentation of the India plate lithosphere
  - 4.4.3. The western and eastern syntaxes
- 4.5. Current seismicity of the Himalaya
  - 4.5.1. Earthquake detection, location and activity
  - 4.5.2. Seismicity of the Himalaya: an incomplete patchwork
  - 4.5.3. Seismicity of the Himalaya: main features
- 4.6. Conclusion
- 4.7. References

Chapter 5. Gravity Observations and Models Along the Himalayan Arc
Rodolphe CATTIN, György HETÉNYI, Théo BERTHET and Jamyang CHOPHEL

- 5.1. Introduction
- 5.2. Methods
  - 5.2.1. Measurements
  - 5.2.2. Corrections
  - 5.2.3. Anomalies
- 5.3. Isostasy
  - 5.3.1. Local compensation
  - 5.3.2. Regional compensation
  - 5.3.3. Effective elastic thickness
- 5.4. Flexure of the Indian plate
  - 5.4.1. Gravity anomaly across the Himalayan belt
  - 5.4.2. Along-strike variation between Nepal and Bhutan
- 5.5. Satellite data contribution

5.5.1. Gravity measurements from space
5.5.2. Towards a three-dimensional image
5.6. Conclusion
5.7. References

Chapter 6. Topographic and Thermochronologic Constraints on the Himalayan Décollement Geometry
Peter A. VAN DER BEEK, Rasmus C. THIEDE, Vineet K. GAHALAUT and Taylor F. SCHILDGEN

6.1. Introduction
6.2. Methods
  6.2.1. Quantitative geomorphic analysis
  6.2.2. Measures of erosion at different timescales: cosmogenic nuclides and thermochronology
  6.2.3. From exhumation to kinematics: thermo-kinematic models
6.3. Regional case studies
6.3.1. Central Himalaya–Nepal
6.4. Discussion
  6.4.1. Constraints on MHT geometry and kinematics at different timescales
  6.4.2. Nature and evolution of ramps on the MHT
  6.4.3. Evidence for out-of-sequence thrusting?
  6.4.4. Lateral segmentation of the MHT
6.5. Conclusion
6.6. References

Part 3. Focus

Chapter 7. Application of Near-surface Geophysical Methods for Imaging Active Faults in the Himalaya
Dowchu DRUKPA, Stéphanie GAUTIER and Rodolphe CATTIN

7.1. Introduction
7.2. Near-surface geophysics
  7.2.1. Geophysical methods for fault mapping
  7.2.2. Case study data and inversion technique
7.3. Geophysical results of case study from south Bhutan
  7.3.1. Electrical resistivity tomography
  7.3.2. Seismic tomography
  7.3.3. Micro-gravity
7.4. Implications of near-surface geophysical findings
  7.4.1. Subsurface imaging
  7.4.2. Overthrusting slip rate assessment
  7.4.3. Deformation at the topographic front

7.5. Conclusion
7.6. References

Chapter 8. Overview of Hydrothermal Systems in the Nepal Himalaya
Frédéric GIRAULT, Christian FRANCE-LANORD, Lok Bijaya ADHIKARI, Bishal Nath UPRETI, Kabi Raj PAUDYAL, Ananta Prasad GAJUREL, Pierre AGRINIER, Rémi LOSNO, Sandeep THAPA, Shashi TAMANG, Sudhan Singh MAHAT, Mukunda BHATTARAI, Bharat Prasad KOIRALA, Ratna Mani GUPTA, Kapil MAHARJAN, Nabin Ghising TAMANG, Hélène BOUQUEREL, Jérôme GAILLARDET, Mathieu DELLINGER, François PREVOT, Carine CHADUTEAU, Thomas RIGAUDIER, Nelly ASSAYAG and Frédéric PERRIER

 8.1. Introduction
 8.2. Measurement methods
  8.2.1. Exploration approach
  8.2.2. Thermal spring water measurements
  8.2.3. CO2 flux and radon flux measurements
  8.2.4. Carbon content and isotopic composition measurements
 8.3. Summary of results at the hydrothermal sites in the Nepal Himalaya
  8.3.1. Overview of hydrothermal sites in Far-Western Nepal
  8.3.2. Overview of hydrothermal sites in Mid-Western Nepal
  8.3.3. Overview of hydrothermal sites in Western Nepal
  8.3.4. Overview of hydrothermal sites in Central Nepal
  8.3.5. Overview of hydrothermal sites in Eastern Nepal
  8.3.6. Overview of hydrothermal sites in the MFT zone
 8.4. Conclusion
 8.5. References

Conclusion
Rodolphe CATTIN and Jean-Luc EPARD

Summary of Volume 3

Tributes
Eduardo GARZANTI, Vincent GODARD, Rodolphe CATTIN, György HETÉNYI, Jean-Luc EPARD and Martin ROBYR

Foreword
Rodolphe CATTIN and Jean-Luc EPARD

Preface. From Research to Education: The Example of the Seismology at School in Nepal Program
György HETÉNYI and Shiba SUBEDI

Part 1. Surface Process

Chapter 1. Orogenesis and Climate
Frédéric FLUTEAU, Delphine TARDIF, Anta-Clarisse SARR, Guillaume LE HIR and Yannick DONNADIEU

    1.1. Introduction
    1.2. Climate in Asia: present and past
        1.2.1. Present-day climate
        1.2.2. Cenozoic climate evolution
    1.3. Reconstructing the paleo-elevation of landforms
    1.4. The contribution of climate modeling
        1.4.1. Impact of orogenesis on the atmospheric circulation
        1.4.2. Impact of orogenesis on the ocean circulation
        1.4.3. Impact of orogenesis on the chemical composition of the atmosphere
    1.5. Conclusion
    1.6. References

Chapter 2. Eroding the Himalaya: Processes, Evolution, Implications
Vincent GODARD, Mikaël ATTAL, Saptarshi DEY, Maarten LUPKER and Rasmus THIEDE

   2.1. Introduction
   2.2. Main process domains in the Himalaya
      2.2.1. Himalayan rivers
      2.2.2. The glaciated High Range
      2.2.3. Critical hillslopes
   2.3. Extreme events and their contribution to denudation
   2.4. 1–10 ka timescale and climatic oscillations
   2.5. Impact of long-term tectonic and climatic evolution over several Ma
   2.6. Out of the Himalaya: sediment transport and storage from the range to the sedimentary basin
   2.7. Conclusion
   2.8. References

Part 2. Natural Hazards

Chapter 3. Glaciers and Glacier Lake Outburst Floods in the Himalaya
Christoff ANDERMANN, Santosh NEPAL, Patrick WAGNON, Georg VEH, Sudan Bikash MAHARJAN, Mohd Farooq AZAM, Fanny BRUN and and Wolfgang SCHWANGHART

   3.1. Introduction
   3.2. Glaciers and their future
      3.2.1. The present state of glaciers at a regional scale
      3.2.2. On the interest of monitoring glaciers
      3.2.3. What will happen to the Himalayan glaciers and Karakoram glaciers?
   3.3. Glacier lakes
      3.3.1. Formation and present distribution of glacier lakes in the Himalaya
      3.3.2. Historic changes of glacier lake abundance and size
      3.3.3. Projections of future glacier lakes
   3.4. Glacial lake outburst floods and downstream propagation
      3.4.1. GLOF triggers
      3.4.2. Dam breach
      3.4.3. Flood propagation
      3.4.4. Early warning
   3.5. Consequences and impact
   3.6. Role in landscape formation
   3.7. Conclusion
   3.8. References

Chapter 4. Landsliding in the Himalaya: Causes and Consequences
Odin MARC, Kaushal GNYAWALI, Wolfgang SCHWANGHART
and Monique FORT

4.1. Introduction
4.2. Understanding landsliding and their links to the dynamics of the Himalayan range
    4.2.1. Preliminary notions on the mechanics of landsliding
    4.2.2. Seasonal landsliding caused by monsoons and extreme rainfall
    4.2.3. Landslide induced by earthquakes and other exceptional perturbations
    4.2.4. Giant and paleo-landslides
4.3. Landslides within Himalayan society
    4.3.1. Hazard cascades and their societal impact
    4.3.2. Human activities as an additional trigger of landslides
    4.3.3. Potential for mitigation
    4.3.4. Climate change and future landsliding
4.4. Conclusion
4.5. References

Chapter 5. Himalayan Surface Rupturing Earthquakes
Laurent BOLLINGER, Matthieu FERRY, Romain LE ROUX-MALLOUF,
Jérôme VAN DER WOERD and Yann KLINGER

5.1. Introduction
5.2. The large devastating earthquakes in the Himalaya
    5.2.1. Historical chronicles and earthquakes
    5.2.2. Strong instrumental earthquakes
5.3. Surface expression of the seismic deformation in the landscape and within paleoseismological excavations
    5.3.1. In the mesoseismal trace of the 1934 earthquake
    5.3.2. In the mesoseismal trace of the 1714 earthquake in Bhutan
    5.3.3. In the mesoseismal trace of the 1950 earthquake
5.4. Overview of the paleoseismological excavations along the Himalayan arc
5.5. Abandoned alluvial terraces, an archive of the paleoearthquakes
5.6. Conclusion
5.7. References

Chapter 6. Seismic Coupling and Hazard Assessment of the Himalaya
Sylvain MICHEL, Victoria STEVENS, Luca DAL ZILIO and Romain JOLIVET

6.1. Introduction
6.2. From current ground motion to the buildup of slip deficit at depth
    6.2.1. Geodetic observations of the interseismic period
    6.2.2. Inferring coupling along the MHT with a Bayesian analysis
    6.2.3. Interseismic coupling distribution

6.2.4. Discussion
6.3. Seismic potential of the MHT
    6.3.1. Conservation of the seismic moment
    6.3.2. Magnitude–frequency distribution in the Himalaya
    6.3.3. Including the physics of fault slip in seismic hazard
    6.3.4. Seismic potential of the MHT
6.4. Seismic hazard in the Himalaya
    6.4.1. Ground motion prediction equations and Vs30
    6.4.2. Modeling scenario events
    6.4.3. Probabilistic seismic hazard assessment
    6.4.4. From hazard to risk
6.5. Conclusion
6.6. References

Part 3. Focus

Chapter 7. Recent and Present Deformation of the Western Himalaya

François JOUANNE, Jean-Louis MUGNIER, Riccardo VASSALLO, Naveed MUNAWAR, Awais AHMED, Adnan Alam AWAN, Manzoor A. MALIK and Ramperu JAYANGONDAPERUMAL

7.1. Introduction
7.2. Structural styles and tectonic prism model
    7.2.1. The main structural zones
    7.2.2. The Himalaya in the west of the syntax
    7.2.3. The Himalaya of the Jammu–Kashmir area
    7.2.4. A structural evolution in agreement with the tectonic prism model
7.3. Out-of-sequence deformation in the western Himalayan syntax
    7.3.1. Quaternary out-of-sequence activity of the NW Himalayan thrusts
    7.3.2. Out-of-sequence seismological and paleoseismological activity in the NW Himalaya
7.4. Deformation associated with a ductile décollement, not always aseismic
    7.4.1. Seismic coupling assessment along the MHT
    7.4.2. Occurrence of ruptures on asperities
    7.4.3. Transition between thin-skin and thick-skin tectonics
7.5. Conclusion
7.6. References

Chapter 8. The 2015 April 25 Gorkha Earthquake
Laurent BOLLINGER, Lok Bijaya ADHIKARI, Jérôme VERGNE, György HETÉNYI and Shiba SUBEDI

8.1. Introduction
8.2. The mainshock and its effects on the ground

8.2.1. Overview description of the mainshock
8.2.2. Ground motion and deformation
8.3. Investigating the seismic source at depth
8.4. Aftershock activity and post-seismic relaxation
   8.4.1. Early aftershocks and the occurrence of the Kodari earthquake
   8.4.2. Seismicity monitored by denser, dedicated temporary networks
   8.4.3. Post-seismic relaxation monitored by geodetic and seismological networks
8.5. A more earthquake-informed and earthquake-resilient local community in the aftermath of the earthquake
8.6. Conclusion
8.7. References

Chapter 9. Crustal Fluids in the Nepal Himalaya and Sensitivity to the Earthquake Cycle

Frédéric GIRAULT, Christian FRANCE-LANORD, Lok Bijaya ADHIKARI, Bishal Nath UPRETI, Kabi Raj PAUDYAL, Ananta Prasad GAJUREL, Pierre AGRINIER, Rémi LOSNO, Chiara GROPPO, Franco ROLFO, Sandeep THAPA, Shashi TAMANG and Frédéric PERRIER

9.1. Introduction
9.2. Overview of thermal springs geochemistry in Nepal
9.3. Overview of gaseous emission zones in Nepal
9.4. Spatial organization of crustal fluid release
9.5. Temporal variations of crustal fluid release: a tectonic control
9.6. Conclusion
9.7. References

Conclusion
Rodolphe CATTIN and Jean-Luc EPARD

Printed by BoD"in Norderstedt, Germany